都市与社区
Community and the City

建筑立场系列丛书 No.41

中文版
(韩语版第357期)

韩国C3出版公社 | 编

时真妹 曹硕 栾一斐 张琳娜 于风军 王莹 时跃 周一 郭薇 | 编

大连理工大学出版社

都市与社区

004 社区复兴调查 _ Andreas Marx

建筑立场系列丛书　No. 41

营造集会空间

008 营造集会空间 _ Tom Van Malderen
012 胜利街居委会 _ Scenic Architecture
022 埃拉索社区中心 _ Romera y Ruiz Arquitectos
028 达普托圣公会教堂礼堂 _ Silvester Fuller
036 内谢尔战争纪念馆 _ SO Architecture

社区的起源：独立个体

046 社区的起源：独立个体 _ Jaap Dawson
052 艾琳娜·加罗文化中心 _ Fernanda Canales + Arquitectura 911sc
064 隆勒索涅城市媒体中心 _ du Besset-Lyon Architectes
074 Daoíz y Velarde文化中心 _ Rafael De La-Hoz Arquitectos
082 Fjelstervang户外社区中心 _ Spektrum Arkitekter
088 友好中心 _ Kashef Mahboob Chowdhury/Urbana
100 新Encants市场 _ b720 Arquitectos

建立社区的情景

112 建立社区的情景 _ Diego Terna
116 La Boiserie多功能活动中心 _ DE-SO Architecture
128 布朗库堡文化中心 _ Mateo Arquitectura
138 Akiha Ward文化中心 _ Chiaki Arai Urban and Architecture Design
152 CIDAM农业经营研发中心 _ Landa Arquitectos
162 艺术广场 _ Brasil Arquitetura
178 比斯开新文化馆及图书馆 _ aq4 Arquitectura

188 建筑师索引

Community and the City

004 *Investigating the Re-birth of the Community _ Andreas Marx*

Building Gathering Spaces

008 *Building Gathering Spaces _ Tom Van Malderen*

012 Victory Street Community Center _ Scenic Architecture

022 El Lasso Community Center _ Romera y Ruiz Arquitectos

028 Dapto Anglican Church Auditorium _ Silvester Fuller

036 Nesher Memorial _ SO Architecture

The Source of Community: Individual Bodies

046 *The Source of Community: Individual Bodies _ Jaap Dawson*

052 Elena Garro Cultural Center _ Fernanda Canales + Arquitectura 911sc

064 Lons-le-Saunier Mediatheque _ du Besset-Lyon Architectes

074 Daoíz y Velarde Cultural Center _ Rafael De La-Hoz Arquitectos

082 Fjelstervang Outdoor Community Hub _ Spektrum Arkitekter

088 Friendship Center _ Kashef Mahboob Chowdhury/Urbana

100 New Encants Market _ b720 Arquitectos

The Scene that Builds a Community

112 *The Scene that Builds a Community _ Diego Terna*

116 La Boiserie _ DE-SO Architecture

128 Cultural Center in Castelo Branco _ Mateo Arquitectura

138 Akiha Ward Cultural Center _ Chiaki Arai Urban and Architecture Design

152 CIDAM _ Landa Arquitectos

162 The Square of Arts _ Brasil Arquitetura

178 New Culture House and Library in Vizcaya _ aq4 Arquitectura

188 Index

都市与社区
Community and the City

胜利街居委会_Victory Street Community Center / Scenic Architecture
埃拉索社区中心_El Lasso Community Center / Romera y Ruiz Arquitectos
达普托圣公会教堂礼堂_Dapto Anglican Church Auditorium / Silvester Fuller
内谢尔战争纪念馆_Nesher Memorial / SO Architecture

营造集会空间_Building Gathering Spaces / Tom Van Malderen

艾琳娜·加罗文化中心_Elena Garro Cultural Center / Fernanda Canales + Arquitectura 911sc
隆勒索涅城市媒体中心_Lons-le-Saunier Mediatheque / du Besset-Lyon Architectes
Daoíz y Velarde文化中心_Daoíz y Velarde Cultural Center / Rafael De La-Hoz Arquitectos
Fjelstervang户外社区中心_Fjelstervang Outdoor Community Hub / Spektrum Arkitekter
友好中心_Friendship Center / Kashef Mahboob Chowdhury / Urbana
新Encants市场_New Encants Market / b720 Arquitectos

社区的起源：独立个体_The Source of Community: Individual Bodies / Jaap Dawson

La Boiserie多功能活动中心_La Boiserie / DE-SO Architecture
布朗库堡文化中心_Cultural Center in Castelo Branco / Mateo Arquitectura
Akiha Ward文化中心_Akiha Ward Cultural Center / Chiaki Arai Urban and Architecture Design
CIDAM农业经营研发中心_CIDAM / Landa Arquitectos
艺术广场_The Square of Arts / Brasil Arquitetura
比斯开新文化馆和图书馆_New Culture House and Library in Vizcaya / aq4 Arquitectura

建立社区的情景_The Scene that Builds a Community / Diego Terna

社区复兴调查
Investigating the Re-birth of the Community

感知实际上不仅仅是感觉。感知是一个积极的过程,通过它我们可以了解周围的世界。

——布莱恩·劳森,《空间的语言》,2001年

本书探讨的是建筑与社区概念的关系。焦点在于从社区的规模上来感知建筑物——也就是使用者以及居住者与周围建筑的互动。首先,简要介绍一下社区的概念以及在社区环境中的建筑设计方法,这将会建立起这种跨学科设计方法的界限,然后看一看与此主题有关的几个项目。

社区的概念在城市研究中经常出现,一般用于参照城市中社会生活的各个方面。传统上,社区这一概念被人类学家、社会学家、地理学家和城市规划者使用,用来指一定的界限、地点或地域内发生的一系列社会关系。在关于城市和社会的研究中,社区是个最具争议的概念。虽然按照惯例,社区的概念是用来描述某一特定地点的特征的,但它也被当成一种更接近意识形态的词语,藉此来证实一种特殊的身份,或是促进一个特殊的政治项目。总体上,可以定义为四种广泛的方法。第一种认为社区是出现在明显空间化的地理环境中的一系列社会关系。从该词语的这个角度来讲,有大量的集中于这种具体社区的形式和功能的著作。第二种方法将社区的概念定义为个体或社会群体的特殊社会互动模式的产物。这种方法的前提是假设存在不同程度的一致以及矛盾,它将社区看成社会成员之间持续协商的产物。第三种意义上的社区用来描述个体与社会之间特殊的社会关系。这个观点和传统意义上对社区的理解最为接近,因为它唤起了社区的传统概念,即对归属感的追求,以及想成为社会群体一员的愿望。第四种方法着眼于社区的基本性质如何随着通信系统和电脑科技使用的变革而得到了彻底的转变。根据这一观点,通信技术的发展从根本上破坏了社区的传统概念,在很大程度上改变了个人和社会群体之间建立归属关系的方式。

"Perception is actually more than just sensation. Perception is an active process through which we make sense of the world around us." - Bryan Lawson, *Language of Space*, 2001.

The issue deals with the relationship of architecture and the idea of community. The focus lies on the perception of buildings at the scale of communities – thus the interaction of users and inhabitants with the surrounding architecture. Before taking a look at several projects related to this topic, a brief introduction on the idea of community and the means of architecture in this context will build up the scope of this interdisciplinary approach.

The concept of community has appeared regularly throughout urban studies and is generally employed in reference to all aspects of the social life of cities. Traditionally used by anthropologists, sociologists, geographers and urban planners to signify a set of social relationships operating within a specific boundary, location, or territory, community is arguably one of the most contested concepts used in the study of the city and society. Although conventionally evoked to describe the characteristics of a specific locality or area, the idea of community has also been used in far more ideological terms as a means by which to substantiate a particular identity or to further a specific political project. In general four broad approaches can be identified. The first approach conceives of community as a set of social relations occurring within a distinctly spatialized and geographical setting. There exists a rich body of work that has focused upon the form and function of specific communities in this sense of the term. A second approach conceptualizes community as the outcome of a particular mode of social interaction among individuals or social groups. Premised upon varying degrees of consensus and conflict, this approach essentially views community to be the product of ongoing negotiation between social actors. Community has been used in a third sense to describe a particular type of social relationship between the individual and society. This perspective is perhaps closest to a commonsense interpretation of community, as it evokes the notion of community as a search for belonging and desire for group membership. The fourth approach looks at how the foundational nature of community has been decisively altered by innovations in the use of communications and computer technology. According to this view, developments in communicative technology have fundamentally undermined more traditional conceptualizations of community and radically altered the means by which individuals and social groups generate bonds of attachment.

关于这一主题的广义范围的调查只能在这一环境中进行探讨。例如，芝加哥大学城市社会学院提出了这样一个城市研究的观点：城市是以"各种小世界拼成"的一种综合的形式。对于一些特殊的社会群体、邻里之间、职业工作（比如土匪、街头混混、出租车司机，及许多其他职业）方面的人类学探索研究，显示出了具有明显公共情绪的特殊的社会文化领地。对于城市增长和发展的分析重心就是人类生态学观点的应用，这一观点认为，持续不断的移民入侵使城市系统持续发展和扩大。芝加哥学院派的关键观点就是认识到了当地和社区层面上的流动性对城市的整体结构有着重要的影响。另外，许多信仰马克思主义的批判型学者论证得出了全球资本主义的崛起对当代城市环境中社区的实现也有不利的影响。他们认为，正是全球资本主义的霸权才导致了城市生活中的社区被取代，这样的观点表明社区与资本世界的秩序是相对立的。这样的讨论多在关于中产阶级化和城市新兴阶级的争辩中进行。并且，国际移民也从根本上改变了当前全球城市中的社区的构成。以维持跨国的个人和经济关系网为特征的当代移民社区有利于全球的社交和科技革新，在移民的祖国和移居国之间形成了联系。

在当代的城市研究中使用社区的概念时，可以看出三个广泛的趋势。第一个趋势包含在关于公民接触的社群主义辩论中。如果对于特定社区的认同依赖于归属感的培养，那么公民结社和当地社区网络的复兴就是必不可少的。将社区的重要性和公民权的重要性结合在一起，就很容易理解社群主义者的立场，他们的立场挑战了个体自主性和个人成就实现方面的自由主义的重要性。

第二种趋势表明，随着越来越多个人因为相同的观点、品味和生活方式而聚集在一起，社区的本质已经成为更加接近唯意志论心理学的社会契约方式。在这种意义上，社区受到了特定的身份或兴趣追求的性质的严格限制，其特征是社区成员的标准具有相对流动性和暂时性的特点。

The broad range of research related to this topic can only be broached in this context. For example, the Chicago School of Urban Sociology proposed a vision of urban study in which the city was revealed as a composite "mosaic of little worlds". Ethnographic explorations of everything from particular social groups, neighborhood locales, and occupational niches(e.g., gangs, street corners, taxi-cab drivers, and many others) revealed distinct socio-cultural enclaves of communal sentiment. Central to that analysis of urban growth and development was the employment of a human ecology perspective in which the city was seen to develop and expand organically due to ongoing waves of immigrant invasion and succession. A key insight of Chicago School scholarship was the recognition that dynamics occurring at the local and community level had significant impact upon the overall structure of the city. Additionally, many critically and Marxist oriented scholars have argued that the rise of global capitalism has proven detrimental to the realization of community in contemporary urban settings. Arguing that the hegemony of global capitalism has led to the displacement of community in city life, such perspectives suggest community to be antithetical to the logic of a capitalist world order. Many such discussions have been conducted within the context of debates on gentrification and urban regeneration. Furthermore, international migration has also fundamentally altered the constitution of community in the contemporary global city. Characterized by the maintenance of transnational personal and economic networks, contemporary immigrant communities avail of the communicative and technological innovations of globalization to forge links between homeland and host country.

Three broad trends can be discerned when considering the utility of the concept of community in contemporary urban studies. A first trend is contained in communitarian debates regarding civic engagement. If identification with a particular community is dependent upon the cultivation of a sense of belonging, the revival of civic associations and local community networks are deemed essential. Combining an emphasis on the importance of community with a stress on citizenship, the communitarian position is most clearly understood when conceived of as a position that challenges the liberal emphasis placed upon individual autonomy and achievement of personal fulfillment.

A second trend suggests that the nature of community has become a far more voluntaristic means of social engagement as individuals come together on the basis of similarity of ideas, taste and lifestyle. In this view, community is heavily circumscribed by the nature of the particular identity or interest pursuit, and characterized by relatively fluid and transient criteria of membership.

第三，因为多媒体和通信科技的发展，社区的社会互动不再局限于当地的地理位置，这从根本上改变了社区构成的方式。从这个意义上讲，社区已经不再依赖于面对面的社会交流，而是更依赖于互动的虚拟网络。

　　从这个关系上来说，建筑设计的方式和角色就必须得到仔细检查了。总体上，建筑设计即指建筑师设计的建成环境，也指这一职业的综合名称。这一基本概念因掺入了很多要素而被复杂化，尤其是"恰如其分的"建筑分类也充满了争议和抗争，就像设计者争取被认可为建筑师的权利一样复杂。将建筑实践设置于特定的城市背景、政治体制和资本模式当中，这种研究使建筑更难于从这些过程中获得自主权。这个职业的立场介于艺术、美学与对有社会意义的形式和空间的创造之间，对后两者更为关注，也关注材料问题的功能反馈，就这一点而言，其职业的立场是重要的考虑因素。例如，关于后现代主义的建筑学讨论主要用于重新引发对于建筑的社会意义的探讨，尤其是开启建筑和空间多重解读的潜力，展示建筑传统风格的多元化。换句话说，建筑的形式对使用者和其他的市民意义重大，是城市空间的标志，也是不同范围的社会实体的反映。同时，聚焦于建筑能够支撑不同社会意义的能力的争论通常是以对不平等的权力关系的批判为代价的，这种权力关系支撑着建筑的社会生产。在本书中列出的项目显示出了在互动的基础上联系社区和建筑的尝试。这一系列项目包括个人规模的（例如，Romera y Ruiz建筑师事务所的埃拉索社区中心）、社区规模的（例如du Besset-Lyon建筑师事务所的隆勒索涅城市媒体中心）和城市规模的（例如千秋新井城市建筑设计事务所的Akiha Ward文化中心）。

Thirdly, the dis-embedding of community from local sites of social interaction, due to developments in media and communicative technology, has decisively altered the means by which community is constituted. In this view, community has become less dependent upon face-to-face forms of social interaction and more upon virtual networks of connectivity.

In this relation, the means and role of architecture have to be scrutinized critically. In general architecture refers both to those parts of the built environment that are designed by architects and the collective designation of the profession. This basic definition is complicated by a number of factors, not least of which is the fact that the types of buildings that can "properly" be considered architecture are of significant controversy and struggle, as are the right of designers to be recognized as architects. Research that situates architectural practice within particular urban contexts, political regimes, and capitalist models problematizes architecture's claims to autonomy from these processes. The profession's position somewhere between an art, primarily concerned with aesthetics and the creation of socially meaningful forms and spaces, and as a primarily functional response to material issues, is an important consideration in this regard. For example, the architectural discourse of postmodernism served to reinvigorate discussion of architecture's social meaning, not least by opening up the potential for multiple readings of buildings and spaces, and by celebrating a plurality of stylistic traditions. In other words, the built forms of architecture become meaningful to users and other citizens as markers of urban space and as reflections of a diverse range of social realities. At the same time, a primary focus on architecture's capacity to support diverse social meanings can often be at the expense of a deeper critique of the unequal power relations that underpin the social production of architecture. The projects outlined in this issue show the attempt to link community and architecture on the basis of interaction. The range includes projects on personal scale (i.e. El Lasso Community Center from Romera y Ruiz Arquitectos), on community scale (i.e. Lons-le-Saunier Mediatheque from du Besset-Lyon Architectes) and on urban scale (i.e. Akiha Ward Cultural Center from Chiaki Arai Urban and Architecture Design). Andreas Marx

营造集会空间
Building Gathering

如果你留意到外界关于建筑以及它与社区关系的众多观点，就会发现到处都是充斥着人们复杂情感的各种各样的激烈讨论。不仅因为"社区"一词本身涉及大量的多维含义，更多的是因为它在建筑中建立了公共机构的职能，而这无关建筑能否解决社会问题。

本章所述的项目以社区设施的形式分布在世界各地，它们都遵从社区的"使命"和期望，是人们集会的首选地点。这种集会行为是几个世纪以来人们一直践行的，集会的人群成为建筑类型多样化的推动力，并促使这种活动发展为一种仪式。社区中心进行着这些活动并不断产生新花样，它们融汇了过去和现在，有助于促进社区的团结，激励人们参与其中，相互认识。同时，全球化使人们开始警觉自己的身份，生怕失去它。本章展示的四个项目证明，地区差异仍然存在，建筑与社区的关系要求建筑行业不仅要注意周边的建设环境，也要重视其内涵。

If we take a look at the numerous opinions out there on architecture and its relationship to community, we can observe that a colorful debate is ongoing and sentiments are very mixed. Not only because the word "community" itself covers a number of meanings and diverse scales, but most certainly because it brings up the role of public authorities in architecture and whether architecture can be a solution to sociological questions or not.

The projects described in this chapter are forms of community facilities, located in different corners of the world. They all come with their community "missions" and aspirations, but they are first and foremost places where people gather. The act of gathering is something we have been doing for centuries and the assembly of people has been a driver for both the development of certain architectural types as well as sets of actions that can be referred to as rituals. Community centers embody these rituals and give birth to new ones. In doing so they blend past and present, help community bonds grow, increase participation and stimulate encounters. Whilst globalization makes us wary of our identity and instils the fear of losing it, the four projects shown here demonstrate that regional otherness still occurs and that the relationship between architecture and the notion of community commands for architectural practices that commit not only to the form of the built environment, but also its content.

胜利街居委会_Victory Street Community Center/Scenic Architecture
埃拉索社区中心_El Lasso Community Center/Romera y Ruiz Arquitectos
达普托圣公会教堂礼堂_Dapto Anglican Church Auditorium/Silvester Fuller
内谢尔战争纪念馆_Nesher Memorial/SO Architecture

营造集会空间_Building Gathering Spaces/Tom Van Malderen

Spaces

建筑与社区概念之间的活跃关系总是能让人抒发大量的情感，表达丰富多样的观点。无论追溯20世纪的大量文献还是浏览近期的文章、博客帖子和社会媒体的大肆宣传，都会发现人们对于建筑之于社区角色和责任的讨论非常激烈。这些观点多种多样，有些积极乐观，有些疑问重重。社区一词的含义可谓包罗万象，从特定群体到国际社会，从物理区域到概念表征、心理归属感。仔细看看本章关于社区建筑的案例，它们以多种组合形式出现，展现了许多新鲜观点，为建筑与社区的关系注入了丰富内涵。

大多数社区建设项目都是由政府当局发起、委任并提供赞助的。这种政府支持的形式与此同时成了争论的焦点，引起社会各界的争议。一方面人们觉得这将形成一个由"开明的"设计师组织建设社区的自上而下的便捷途径，而在这个过程中，这些政府赞助的工程就会被指绕过了所有的参与流程和社区决策框架。另一方面，私人开发商门前也一直有游说团体在反对政府当局赞助社区建设。他们指控这是社交工程的尝试，并称之为集体主义思想的残余，虽然这种思想在几十年前曾风靡一时，但通常以失败告终。

建筑是否能够或应该解决社会问题，将成为长期存在的争议。曾经有段时间人们不再认为社区是社会的产物，越来越多的人把它当成社会的特定部分，其中新兴的封闭式社区或许是最让人担忧的变化。这类社区在真正意义上只注重私人空间，对社区空间毫无兴趣，更不用说创造集会或偶遇的可能了。

集会空间是本章将要仔细讨论的四个项目的共同特点。埃拉索社区中心和内谢尔战争纪念馆是和风景融为一体的户外集会场所；达普托圣公会教堂礼堂利用一些室内空间组织聚会；而胜利街居委会则是交织而成的一个集群空间。

当然，这些中心也有其基本的公共议程、任务或期望，虽然给人一种超定的第一印象，但它们却是集会场所悠久传统的一部分。启动这些项目的政府必然希望这些场所能够掩盖社会丑恶现象、对抗异化、减少贫困、抑制公共暴力。然而，其最初动机不是要改变人们的行为，或强调一种趋势，而是使这些场所成为人们长久以来习惯的聚会仪式的基础设

The relationship between architecture and the notion of community has always been a very animated one, bringing up bags of sentiments and diverse opinions. Whether we trace back in time to the many 20th century manifestos or look for more recent essays, blogs posts or social media blurbs, we find a very rich and active debate about architectures' role, promise and responsibility towards community. The opinions out there are very diverse and the tone ranges from the very optimistic to the very problematic. The meaning of the word community itself covers a wide variety of scales and levels, from identifying a very specific group to the idea of an all-encompassing world community, and from referring to a physically defined area to a conceptual representation or psychological sense of community. When taking a closer look at the built examples of community architecture in this chapter, these multiple levels and scales come to the foreground in a variety of combinations, and add to the rich relationship between architecture and the notion of community and the many opinions towards it.

Most community building projects are initiated, commissioned and sponsored by a form of public authority. This support by a form of authority is at the same time also a point of contention that raises criticism from all ends of the spectrum. On the one hand the argument is being raised that it forms part of a top-down approach with the "enlightened" architects in charge of helping the community through their work. And in doing so these authority sponsored works are criticized of bypassing any participatory processes or community-based decision framework. On the other hand there is the ongoing lobby from the private developers' front against the same public sponsored community architecture. They accuse it of being social engineering attempts and call it a remainder of the large state-collectivist ideas that were in fashion up to a few decades ago and often failed.

It is continuous to be one of the great debates whether architecture can or should solve sociological issues or not. Certainly at a time where the idea of "community" is less and less considered to be a product of society; but instead increasingly used to refer to very specific parts of that same society. Amongst these parts the rising popularity of the gated community is probably the most uneasy one. Paradoxically, this is a community that is predominant about private space with little to no interest towards community space in the real sense, let alone space for gathering or accidental encounters.

Space for gathering is something that the four projects we will take a closer look at in this chapter all have in common. El Lasso Community Center and Nesher Memorial have outdoor gathering spaces carved out of the landscape. Dapto Anglican Church Auditorium organizes the gathering through a set of indoor spaces, whilst Victory Street Community Center stitches and weaves a whole cluster of gathering spaces together.

Of course these centers also have their underlying "public" agen-

施。这些设施将回溯如希腊集会这样的古建筑,甚至超越它们。

集会的各项活动要根据仪式按固定顺序进行,仪式是大多数国家和文化的特色之一。与季节、灯光、地势、材料、工艺和场地一样,建筑通常在仪式当中担当活性剂的角色。现如今的仪式也许和过去极其相似,只是隐约在人们的行为中有所体现,甚至被忽略掉。不过,它们扩大了物理空间,为建筑内涵提供了更深一层的意义。所以,不可否认建筑可以诞生当代的仪式。

位于格拉纳达的埃拉索社区中心的建筑风格反映出许多种文化符号,无论是古老的还是新兴的仪式,都有所关联。为了在建筑物和大海之间设计集会场所,建筑师拆毁了岛屿地形的内部花园,这类似于祖辈们为了种植而留出一片土地,修建挡风墙以防止水土流失。虽然Romera y Ruiz建筑师事务所在其作品中通常用色谨慎,不过将对这个户外社区花园的立面大胆利用彩色。这样的设计如同彩虹,迎合了现代社会宽容、平等的象征。

本章所提到的几个项目都体现了过去和现在的融合。社区的黏合度需要花时间去磨合,同样建筑本身也需要时间得到关爱,并赋予社区一种自尊和骄傲。这就是建筑的美之所在,一种视觉形式开始有所贡献,可见审美水平和物理品质对于社会工程的重要价值不可否认。建筑能够提供一个物理空间,而社区通常希望这个物理空间能得到社会的认可。

Silvester Fuller建筑师事务所在设计达普托圣公会教堂礼堂时,这里没有足够的实际空间。一旦新建筑周边的建筑体量得到了确定,那么流通空间就可以零零散散地分布在其中。连接着礼堂、现存教堂和幼儿园大楼的流通区域在宽敞与紧密之间保持一种微妙的平衡,成为一个和谐的社交场所,里面一部分区域朝向户外风景,一部分则完全封闭。建筑体量和空隙看上去就像海绵状组织,不断与周围的环境进行交流。

由山水秀建筑事务所设计的胜利街居委会的流通布局成功地体现了微观城市规模的理念。该项目位于上海市青浦区的一个水乡,利用了将流通场所交织在一起的传统方法。这不仅是为了完成此历史保护区的设计规划,也是希望和附近的古建筑开启对话。建筑师沉溺于一种特别的外形设计当中,使灯火闪耀、静谧无声的庭院有编织交错的感觉。他们

das, missions or expectations and can even come across as a little over-determined at first glance, but they all form part of a long tradition of building gathering spaces. The authorities that started off these projects may sure hope these centers mask certain shortcomings within their society, fight alienation, make poverty bearable or suppress public violence. However, their first motive is not to be found in changing behavior, nor in addressing a passing trend but in being part of infrastructures that cater for a centuries-old habit or ritual of gathering. These infrastructures go back to longstanding constructs like the Greek agora and way beyond.

Gathering holds a close affinity to the sequence of activities or sets of actions and movements that are inherent to rituals. Rituals feature in most societies and cultures. Together with seasons, light, topography, material, crafts and occupation, architecture has always been an active agent in the process of ritual. Nowadays, rituals may feel very much like something from the past and only faintly resonate in our actions or even go unnoticed. Nonetheless they expand the physical space and can provide a deeper layer of meaning to a building. In fact, we shouldn't exclude that architecture can give birth to contemporary rituals.

El Lasso Community Center in Granada reflects many cultural identifiers that relate to rituals, both old and new ones. For the creation of the gathering space between the building and the sea, the architects scraped the inner garden out of the islands' topography, similar to the way their ancestors prepared the fields for cultivation and protected them with stone walls to break the wind and avoid erosion. And although Romera y Ruiz Arquitectos often employ selected colors in their work, for the facade bordering the outdoor community garden they chose to incorporate a play on all the colors. In doing so, it reflects a rainbow and echoes the contemporary symbol for social tolerance and equality.

The presented projects all make use of a dialogue between past and present. Community bonds will take time to grow, the same way it will take time for the building to be loved and contribute to the self-esteem and pride of a neighborhood. This is where the beauty of a building, its visual form starts to contribute. Aesthetic and physical quality are important values to a social projects that shouldn't be forgotten. Architecture can offer a physical space and a group or community often needs a physical space to be recognized in a society.

There was not much physical space available to Silvester Fuller Architects when they were invited to design Dapto Anglican Church Auditorium. Once the perimeter mass of the new building was defined, circulation spaces were carved out of the mass. Circulation connects the auditorium with the existing church and preschool building to create an integrated community space with a delicate balance between generosity and intimacy; with some spaces open to the landscape and others completely sealed from it. The masses and voids become like a spongy tissue, defined in a continuous exchange with its surroundings.

Also in Victory Street Community Center, designed by Scenic Architecture the circulation layout takes up a crucial role on the mi-

所追求的不是在其他项目上已经取得成功的现代方法，而是那些几乎要消失的社区中的紧凑结构。用建筑师的话说："山水秀建筑事务所坚信建筑的精髓在于人们怎样看待自然与生存的基础。我们通过建筑来探索空间和时间如何相通相融，以及怎样才能在人类、自然和社会三者之间建立起和谐有机的关系。"

社区中心组织利用实际空间的另一个方式就是通过平衡地区间的集会场所分布而设计私人空间。公共场所和私人空间的衔接促进了人们在社区的相遇和交流，这种民主大众化的场所意味着为彼此陌生的人们创建了一个公开交流的论坛。与私人和公共场所的特色与规模之间的交融以及过渡对社区建筑来说正是至关重要的。

内谢尔战争纪念馆是由政府邀请SO Architecture建筑公司精心设计的，全年都可以聚会。建筑师在馆内建了一个礼堂，供公共演讲、青少年活动和电影放映时使用，再添加一个观景平台和一个看台，连接了社区的儿童游乐场，一改往日附近广场的安静场面。在这个集会场所中有一个极易使人陷入沉思的空间，叫"记忆箱"。这里的窗户给人一种亲密感，人们更愿意把这里作为私人空间进行冥思和怀念。

严格地说，这里介绍的四个项目的设计工作并没有经过包括小型社区联盟、公共机构、非正式工作团队和个体讨论在内的完整的参与过程，但是它们致力于使建筑最终能体现出交流性品质。这四个项目也证明了地区差异仍然存在，并影响着社区组织设计的方式，以及所选择的外形和材料。近来，有人议论设计师只负责建筑环境而不管内部设计。这些项目正是成功的范例，向人们证明建筑师并没有不负责任地忽视项目将带来的影响。而且当人们探求建筑与社区之间的关系时，无疑会发现其中蕴含的建筑与政治的关系。

cro urban scale. This project is located in a water-town in Qingpu, Shanghai, and makes use of the traditional approach of lacing circulations together, not only to address the planning regulations of this historical preservation district, but to reassure a dialogue with the aged neighboring buildings. The architects indulged in a configuration of threading and knitting the courtyard spaces with the rhythm of light and silence. They did not pursue a contemporary approach which they successfully apply in their other projects but the tight and intimate structure of these quickly disappearing communities. To quote the architects: *"Scenic Architecture believes that the spirit of architecture exists in how people perceive the basics of nature and living. We use architecture to explore how space and time stimulate and absorb each other, and how to establish balanced and dynamic relevance among human, nature and society."*

Another way the community centers organize the physical space is by balancing the gathering spaces with areas designed for personal space. The transitions between the communal or public spaces and more private spaces encourages encounters and strengthens communities. Democratic space means creating a forum for "strangers" to interact. Playing with the character and scale of these private and public spaces and the transition between them are essential to community buildings.

In Nesher, SO Architecture was invited by the municipality to rethink a memorial place, and give the place a meaning for gathering throughout the whole year. The architects included an auditorium that runs a public program of lectures, youth events and film screenings, and transformed the adjacent square by adding a viewing platform, a grandstand and improving the connection with the neighbourhood children's playground. Amidst the gathering spaces there is also a more contemplative space called the Memory Box. The windows used for this memory space produce a sense of intimacy and provide for a more personal space to reflect and commemorate.

So whilst the four projects shown here, strictly speaking, haven't been designed by means of a full participatory process including small neighbourhood associations, institutions, informal working teams and individuals themselves; the built results all engage participation in their use and show proof of communicative qualities in their architecture. The four projects also demonstrate that regional differences continue to exist, and that regional otherness and identity influence the manner in which the communities are organized, form finds its shape and materials are applied. Only recently, it was argued that architects are in charge of the form of the built environment, not its content. These projects show excellent examples from architectural practices that don't try to absolve themselves of the responsibility for the full range of implications of their projects. When looking for the relationship between architecture and the notion of community, one undoubtedly comes across the relationship between architecture and politics.

Tom Van Malderen

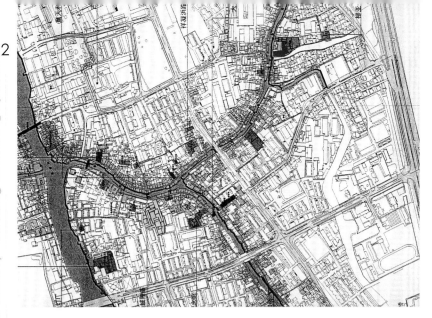

胜利街居委会
Scenic Architecture

胜利街地处上海朱家角古镇的东南一隅，这是一片安静的住宅区，远离北部喧嚣的观光旅游景区。这所新建的街道居委会建筑坐落在两条河流的"丁"字交汇处，将为当地居民提供社区办公室、网吧、阅览室、娱乐室、茶室、健身室以及老年人日托站。

当地城市规划部门提出了几点要求，其中之一是由于建筑基地位于历史风貌保护区之内，因此新建筑在尺度和风格上都要与周围的传统建筑保持一致。于是，事务所决定遵循传统的木结构系统，并委托更加熟悉这种体系的本地承包商来负责工程的施工部分。这样的决定让设计师从一开始就放弃了追求"创新的形式"这一想法。

江南（长江以南）地区住宅楼的传统体系源于清朝的姚承祖所著的《营造法原》（建筑的原理和起源）一书，设计师将这本书用作设计指引，并制定了以下原则：

1. 建造这座建筑的目的是为了丰富本地社区的大众生活。设计师希望通过本地居民所熟悉的建筑语言来巩固其地方文化认同感，为本地居民提供安全和舒适的生活。

2. 沿袭《营造法原》在结构和基本构造上的传统，同时除去多余的修饰。西立面以木质幕墙代替了通常使用的实心侧壁，通过相对半透明的方式照亮传统的结构框架，同时也向公众呈现了"新式现代公共建筑"的姿态。

3. 庭院是整座建筑的灵魂。每个庭院的氛围都来自于空间规模、新建筑物、既有的相邻建筑、阳光、风和时间。

4. 建筑的有机体是一系列内外空间交替的集合体验。

根据以上规则的指引，设计师编织出了一组院落建筑，着迷于穿针引线的布局、晦明变化的庭院空间以及同邻里老屋的直接对话。设计师欣然地享受着个中乐趣，而无意再追逐新鲜的建筑样式。

由北门进入屋内，中间一条走廊经过两侧的社区办公室和网吧，面向前庭敞开着。这个庭院朝向东侧一所原有的旧宅边墙，同时面向门外西边优美的景区敞开着，带有一个容纳了阅览室和餐厅的多功能休息区，环绕着南部的一个中心院落，两侧的走廊容纳了茶室、麻将室和会议室。中央庭院借景东边的传统民居，通过一条残道与西南入口相接。这条路还通过一个东南向的采光天井中转通往老年人日托站。中心的建筑物巧妙地将自己向东折叠起来，面向南部的一个狭长的庭院敞开，原有的葡萄植物爬满了院落古旧的围墙，朝向一个小采光天井，天井有几扇高大的面北的长窗，在保证通风换气的同时也保护了后面原有住宅的私密性。

建筑物的使用者差不多要在其中度过一整天的时间。设计师相信不同规模的房间、走廊和大小不一的庭院，以及浮现其中的旧宅及景致时会让置身其中的人们感知到时间的流逝和静谧。建筑形制的单一并不妨碍空间在形和意上的变化万千，新既出于虚，又何在乎实耶？

Victory Street Community Center

Victory Street is located at the southeast area of Shanghai Zhujiajiao ancient town which is a calm residential neighborhood away from the touristic area to the north. Located at a "T" cross between two small rivers, this newly built community station will serve the local residence with community committee offices, internet bar, reading room, recreation, teahouse, exercise space and oldies' daily care station.

As one of the requests from the local urban planning authorities, the new building needs to follow the surrounding traditional buildings in terms of the scale and style since it is within the historical preservation zone. We decided to design the project with the traditional basic wooden building system, and execute the construction by local contractors who are familiar with the system. This decision kept us away from looking after "creative forms" from the beginning.

The traditional system of Jiangnan (South of Yangtze River) residential buildings origins from the book *Ying Zao Fa Yuan* (*Principles and Origins of Building*) by Yao Chengzu from Qing Dynasty. We used the book as our design guide and set up the following rules:
1. The objective is to provide a place for local community's public life. We hope the place can provide safety and comforts with local

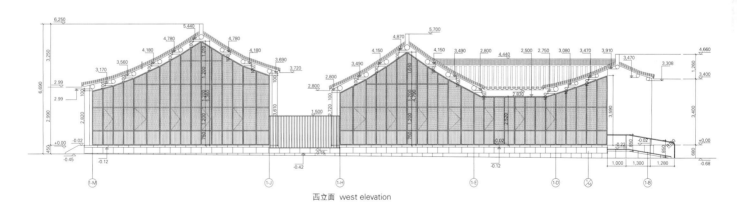

西立面 west elevation

A-A' 剖面图 section A-A'

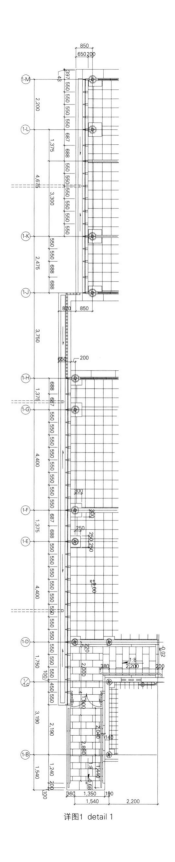

详图1 detail 1

1 办公室	1. office
2 信息中心	2. information center
3 档案室	3. archives
4 医务室	4. infirmary
5 阅览室	5. reading room
6 食堂/会议室	6. canteen/meeting room
7 厨房	7. kitchen
8 棋牌室	8. chess/card room
9 盥洗室	9. wash room
10 会客室	10. reception room
11 茶室	11. tea room
12 健身室	12. gymnasium
13 日托站	13. day care center
14 储藏室	14. store room
15 门廊	15. porch
16 户外活动区	16. outdoor activities

cultural identities established by architectural language that the local residence are familiar with.

2. Follow the tradition from *Ying Zao Fa Yuan* on structure and basic tectonics, while eliminate decorative elements. Change the usual solid sidewall on the west elevation to a wood curtain wall to illuminate the traditional structural frame and present a gesture of a "new public building" with a relatively translucent method.

3. The courtyard is the soul. The aura of each courtyard comes from spatial scale, new building, existing neighbor, sunlight, wind, and time.

4. The architectural organism is a collective experience of a series of alternate interior and exterior spaces.

Guided by the above rules, we organize a group of courtyard buildings. We indulged ourselves in acting as a go-between to connect different courtyard spaces and to chat with the old neighbors. We, as architects, felt a spontaneous enjoyment in this process without any desire to pursue any fancy new forms.

Accessing the building from the northern gate, a middle corridor passes the community office and internet bar on both sides and opens to the front courtyard. Facing the sidewall of an existing old house to the east and opening to the outside scenic context to the west, this courtyard introduces the multi-use lounge that houses the reading room and dinning spaces, which surrounds a central courtyard to the south with wing corridors containing teahouse, mahjong and meeting rooms. The central courtyard borrows the traditional house from the east side as a scenery and connects the southwestern entry with handicapped ramp. It also leads towards the oldies' daily care station with a transferring light well to the southeast. The station building subtly folds itself to the east and opens to a long and narrow courtyard in the south with an old wall full of existing creeper, and to a small light well with long high windows to the north for the breeze ventilation while keeping the privacy from the existing houses behind.

The users of the building would more or less spend a whole day inside. We believe that the rooms, corridors, courtyards in different scales and the emerging old houses and sceneries from time to time will have them perceive the flow and calmness of the time. The monolithic building typology does not necessarily obstruct the freedom of space. Solid form never minds when the freshness comes from void. Scenic Architecture

项目名称：Victory Street Community Center
地点：Zhujiajiao, Shanghai, China
建筑师：Scenic Architecture
合作方：ACO Architects & Consultants PTE LTD
结构和机械工程师：Chen Xiaohui
甲方：Zhujiajiao Town Government
功能：community services
用地面积：663m²
总建筑面积：502m²
竣工时间：2011
摄影师：courtesy of the architect

埃拉索社区中心
Romera y Ruiz Arquitectos

建筑项目始终要与所处的地域特性进行交涉,而交涉的最终结果有时是偏向于我们的。埃拉索社区中心位于西班牙拉斯帕尔玛斯港口附近,一向疏于管理。原先的街道和山谷地势限定了社区的形态及空间,因而该项目将自己限定其中,在一座公园和一处观海点之间制造邂逅。

建筑物常常构筑了这些不规则的边界,就像是在山谷中建造墙体以避免建筑被泥沙冲蚀。创造一座场所的欲望永远优先于对于建筑成就的追求。

内部花园构筑了一个由弧形的围墙、建筑形式及树影构筑成的开放的公共区域。这里成了描绘公有社会的场景。在海景和花园之间,水平的建成阶地面向大西洋,充当了一个过滤器,过滤了光线、空间、从北部吹拂而来的微风和景观。

在某一特定时刻,生动的色彩和投影洒满了俯瞰着海洋的立面,将立面变成了一幅如极光一般的风景画。因而,外立面成了主角,而非一个站着不动供人观赏的无关紧要的小角色。在平面设计上,建筑形式是一堵厚重的围墙,空间被精心雕琢于其上并选好了朝向,从而时刻捕捉外界的风光。建筑小心地处理与斜坡地势的关系,在不同的水平高度设置了两个入口。

这座几乎从农耕中解脱出来的建筑成了加那利群岛风光的化身,色彩表现力、干燥的泥墙、奔流和排水沟充斥着整个画面。

El Lasso Community Center

The architectural project always negotiates with the territory, although sometimes this negotiation comes to us. At the El Lasso Community Center, a neglected neighborhood of Las Palmas, the existing street and valley topography define a form and a space,

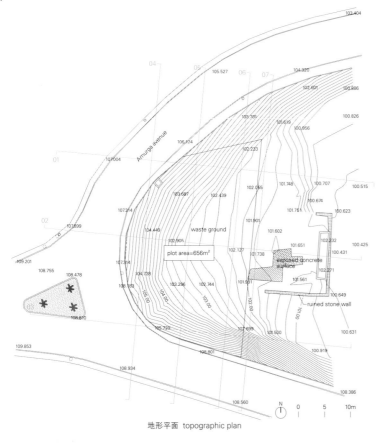

地形平面 topographic plan

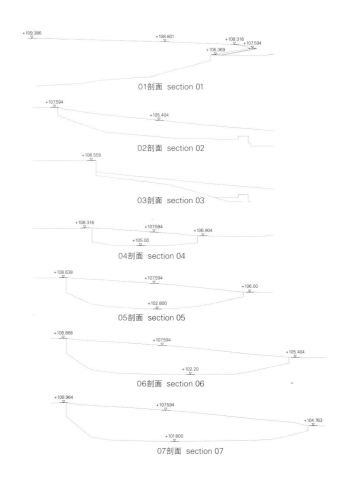

01剖面 section 01
02剖面 section 02
03剖面 section 03
04剖面 section 04
05剖面 section 05
06剖面 section 06
07剖面 section 07

项目名称：El Lasso Community Center
地点：Avenida Amurga, El Lasso, Las Palmas de Gran Canaria, España
建筑师：Pedro Romera García, Ángela Ruiz Martínez
开发商：Excmo. Ayuntamiento de Las Palmas de Gran Canaria
技术建筑师：Manuel Hernández Vera
合作方：Jorge Hernández Fernández, Rocío Narbona Flores and Paula Cabrera Fry
绘图员：Gwendolyn Méndez López
顾问：CQ Ingenieros y Asociados
面积：498m²
竣工时间：2010
摄影师：courtesy of the architect

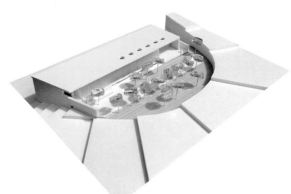

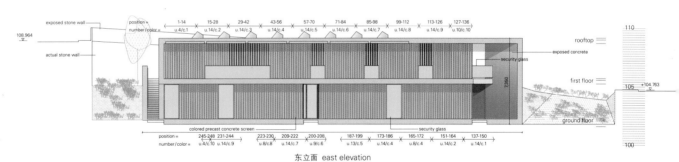

东立面 east elevation

彩色预制混凝土屏幕比色卡
colored precast concrete screen color chart

NUMBER	PAL CODE	DESCRIPTION	POSITION SCREEN NUMBER	TOTAL
1	L079-Y	oro	1-14, 137-150, 249-261 y 262-277	57
2	L058-B	verde pistacho	15-28, 151-164	28
3	L052-C	verde árbol	29-42, 165-172	22
4	K038-B	turquesa	43-56, 173-186	28
5	K030-C	azul petróleo	57-70, 187-199	27
6	L017-C	azul marino	71-84, 200-208	23
7	L148-C	violeta oscuro	85-98, 209-222	28
8	L145-B	violeta claro	99-112, 223-230	22
9	L141-C	granate	113-126, 231-244	28
10	Y030-C	marrón	127-136, 245-248	14
				277

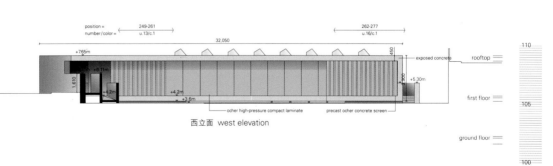

西立面 west elevation

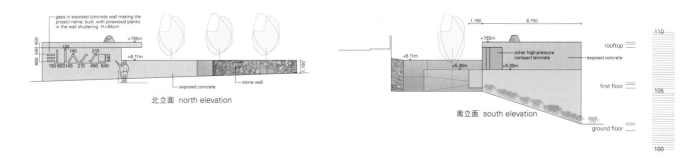

北立面 north elevation　　　　　南立面 south elevation

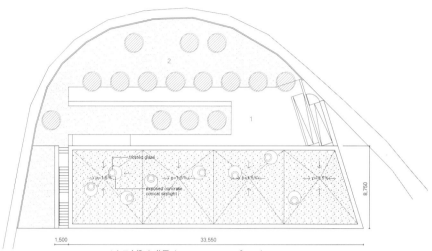

1 入口广场 2 花园 1. entrance square 2. garden
屋顶 roof

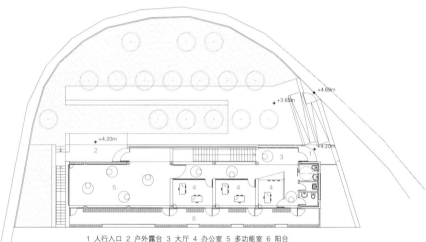

1 人行入口 2 户外露台 3 大厅 4 办公室 5 多功能室 6 阳台
1. pedestrian access 2. outdoor terrace 3. hall 4. office 5. multipurpose room 6. balcony
二层 second floor

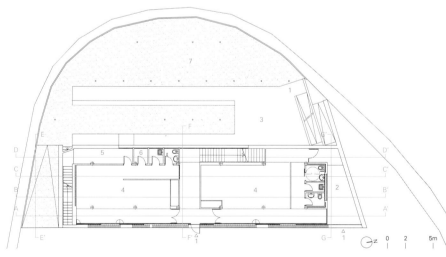

1 人行入口 2 入口大厅 3 入口广场 4 多功能室 5 储藏室 6 衣帽间 7 花园
1. pedestrian access 2. access hall 3. entrance square
4. multipurpose room 5. storage room 6. locker room 7. garden
一层 first floor

and the project just limits itself to draw limits, to construct the encounter between a garden and a viewpoint towards the sea. Often the architecture constructed these permissive borders, like the walls raised in valleys to avoid their own erosion. The desire to create a place is always prior to the architectural accomplishment. The inner garden invents an open collective dominion formed by an enclosing curved wall, the built form and the tree shades. It becomes the scenario for representation of the communitarian use. In between the sea-views and the garden, the longitudinal built terrace watches the Atlantic Ocean, acting as a filter of light, spaces, northern breezes and contained views.

Versus time, the alive colors and casted shadows flood the facade that overlooks the sea, turning it into a landscape of polarized light. The facade turns into an actor rather than a passive element that just stands and shows. On plan, the built form is a heavy wall on which spaces are carved and oriented to catch snapshots of the exterior. The building section gently negotiates the sloping topography allowing for two accesses at different levels.
Architecture almost with the ease of farming becomes content of the territory and incarnation of the Canarian landscape, full of color, dry mud walls, torrents and drains.

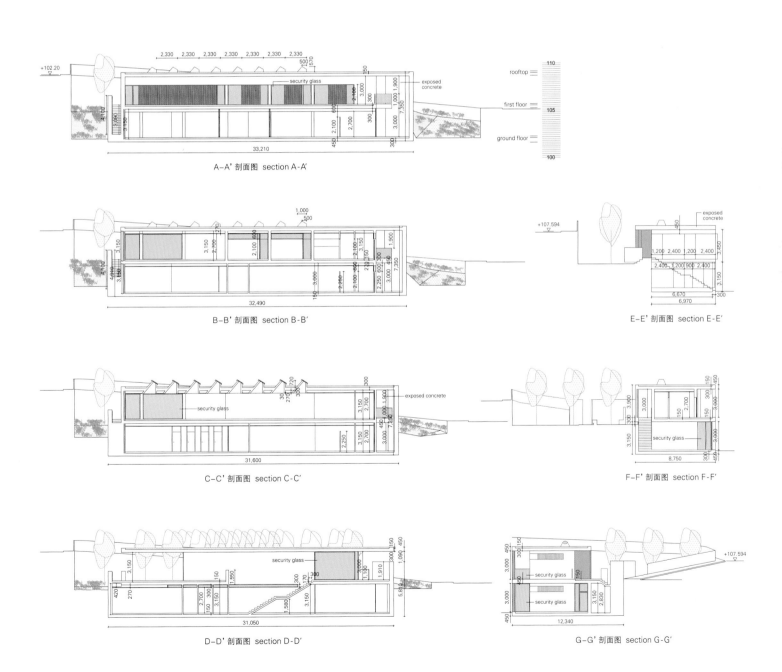

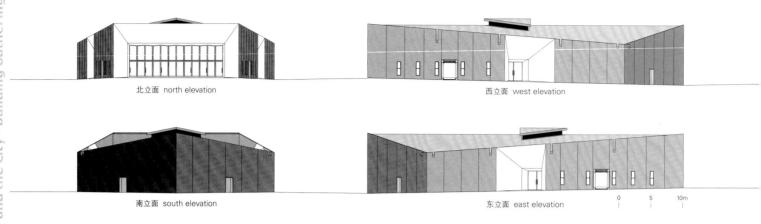

北立面 north elevation　　西立面 west elevation

南立面 south elevation　　东立面 east elevation

达普托圣公会教堂礼堂
Silvester Fuller

Silvester Fuller设计的达普托圣公会教堂礼堂是首个为达普托的圣公会教区设计的新生代建筑。该设计是为了顺应教堂正在转变的功能性和社会性的趋势及其与社区的关系而产生的。作为附近的圣卢克小礼拜堂的补充，该礼堂提供了一个类似于剧院的场所以便应用于更广泛的活动种类。新教堂不再仅仅服务于主日礼拜活动，还要一周七天，每天早、中、晚均用于举办一系列活动，迎合当地社区的广泛需求。

建筑基地的组织策略包括重新布置场所周边的车辆交通，预留出了一片车辆禁行的步行中心。随后新礼堂才会被放置在场所内，避免打扰到既有建筑物。基于这个原因，新礼堂周边的平面设计受到了两座原有建筑——一所幼儿园和一座教区大厅的限制。安置在这两座设施之间使这座新建筑有机会成为一个中央枢纽，新旧建筑内所有重要的活动场地都能够通过这里连通。这个中央枢纽也成了校园聚会的场所。

一旦确定了新建筑的周边小建筑，循环通道空间就会被切分在这些体量之外，再随着人流从停车场走向建筑物，在建筑内部及两个主要空间——礼堂和门厅的周边行走而确定通道的位置。建筑体量的减法设计形成了多个空隙，空隙将这些空间彼此以及与周边景致连接起来。于是主要的支持空间占据了其余的厚实体量。对个体空间的需求要求在开放性和私密性之间达到一种微妙的平衡，因此有些空间向周边景致敞开，而有些则完全遮掩了起来。

外立面的设计顺应了两种现实条件：立面的表面是黑色的，像土一样而且手感粗糙。与之形成反差的是，凹陷进去的区域表面是明亮、平滑、整洁的，标示了建筑的入口并充当了收集设施。一旦走进建筑，进入主礼堂的入口是一个与外部完全逆转的空间，嵌入式的暗黑无光的开口充当了门户，这里通向一个500座的大剧院。剧院是一个仅仅聚焦于舞台的黑盒子，这里预留了一个位置用于日后在舞台上方建造一个自然发光的灯罩。

有限的预算要求构造简洁以及空间透明，并得到有效利用，以此来创造一座易于理解同时又脱离于其周围环境而独立存在的建筑。新建筑旨在创建一种新的设计趋势，关注教区民众，这也是总体发展规划的第一个阶段。

Dapto Anglican Church Auditorium

Silvester Fuller's Dapto Anglican Church Auditorium is the first of a new generation of buildings for the Anglican Parish of Dapto. The design is a response to the changing functional and social direction of the church and its relationship with the community. Intended to complement nearby St Luke's Chapel, the auditorium offers a theater-like venue for a broader range of event types. No longer being a place devoted solely to Sunday worship services, the new church building is required to support a range of events held in the morning, afternoon and evening, 7 days a week and catering to a broad spectrum of the local community.

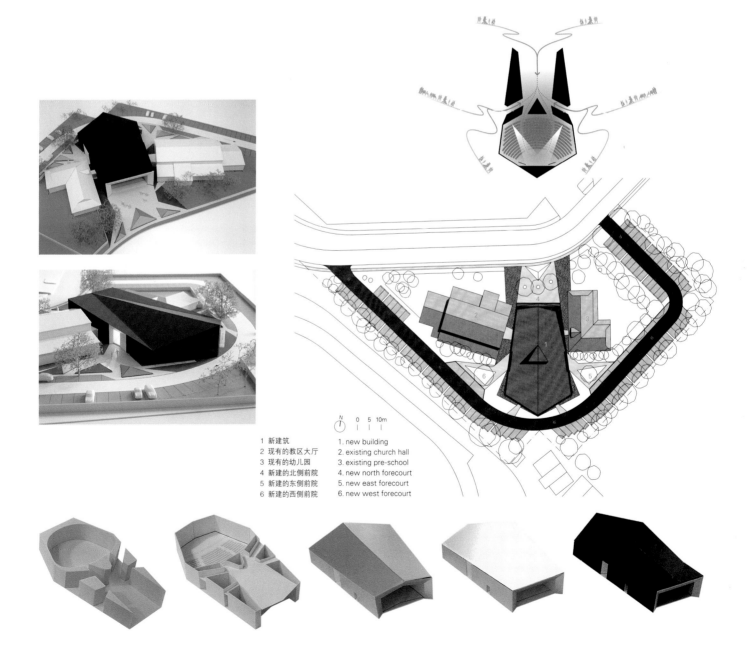

1. 新建筑 / new building
2. 现有的教区大厅 / existing church hall
3. 现有的幼儿园 / existing pre-school
4. 新建的北侧前院 / new north forecourt
5. 新建的东侧前院 / new east forecourt
6. 新建的西侧前院 / new west forecourt

项目名称：Dapto Anglican Church Auditorium
地点：Dapto, NSW, Australia
建筑师：Silvester Fuller
项目团队：Patrik Braun, Rachid Andary, Bruce Feng
项目负责人：Jad Silvester, Penny Fuller
项目经理：Heymann Cohen
结构工程师：Simpson Design Associates / 水力工程师：Whipps Wood
环境与立面工程师：ARUP / BCA顾问：BCALogic
施工技术员：Heymann Cohen / 施工顾问：Think Projects
总承包商：Premier Building Group
甲方：Anglican Parish of Dapto & Anglican Church Property Trust
用地面积：9,546m² / 新建筑面积：1,155m²
设计时间：2008—2009 / 施工时间：2010—2012
摄影师：©Martin van der Wal(courtesy of the architect)

The organizational strategy for the site involved the relocation of vehicular traffic to the site perimeter, allowing for a fully pedestrianized center. The new auditorium was then to be located on the site with minimal intervention to the existing buildings. For this reason the perimeter plan of the new auditorium is bounded by the two existing buildings: a preschool and a church hall. Locating the auditorium between these two facilities presented the opportunity to create a central hub, from which all the primary event spaces, in both the new and existing buildings are accessed. This hub becomes the campus' meeting place.

Once the perimeter mass of the new building was defined, circulation spaces were carved out of the mass, informed by the flow of people from the parking areas to the building and subsequently in and around the two primary spaces: the auditorium and foyer. This subtraction of mass defines voids which connect these spaces to each other and the landscape. The secondary support spaces then occupy the remaining solid mass. The requirements of the individual spaces called for a delicate balance between generosity and intimacy, with some spaces open to the landscape and others completely concealed from it.

The external facade responds to two conditions: where the primary mass has been retained the facade surface is dark, earth-like and roughly textured. In contrast the subtracted void areas are bright, smooth and crisp surfaces identifying the building's entrances and acting as collection devices. Once inside the building, the entry into the main auditorium is an inverse of the exterior, presenting recessed darkened apertures acting as portals which then open into the 500 seat theater. The theater is a black-box with a singular focus on the stage. There is the provision for a natural-light-emitting lampshade to be built above the stage at a later date.

A modest budget demanded construction simplicity combined with spatial clarity and efficiency, to produce a building that is easily understood whilst standing apart from its context. The new building aims to establish a new design direction and focus for the Parish and is envisaged as stage one of a master plan of growth.

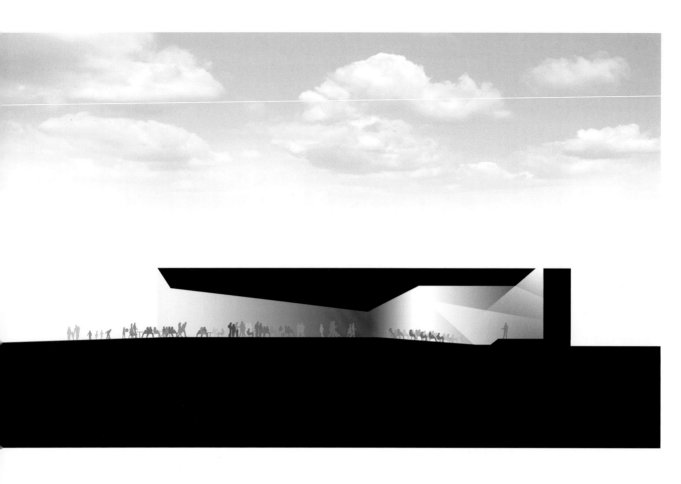

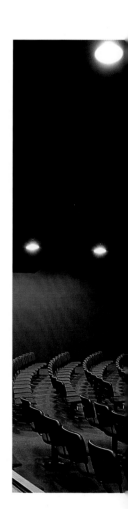

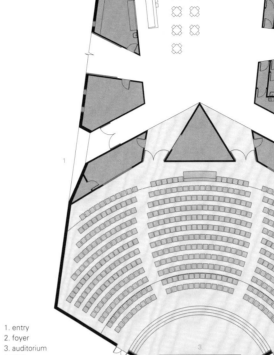

1 入口处	1. entry
2 门厅	2. foyer
3 礼堂	3. auditorium
4 卫生间	4. WC
5 婴儿换巾处	5. baby change
6 啼哭室	6. crying room
7 护理室	7. nursing room
8 储藏室	8. store

一层 ground floor

以色列内谢尔战争纪念馆建于原先的历史保护建筑的基础之上，该建筑在过去是作为防御位置使用的。在1948年的阿拉伯-以色列战争中，这一位置曾用来保护内谢尔的居民。这座建筑位于一个安静街区的心脏部位，在一座小山顶上，被一个巨大的广场和一个儿童游乐场所环绕。

在施工之前，这座建筑疏于管理，一年仅一次用于公众活动——全国阵亡将士纪念日，其余时间则向游客开放，展示保存公众记忆的照片，作为民众纪念阵亡将士仪式的一部分。

内谢尔市政当局委托SO建筑事务所来扩建环绕建筑的广场，使它能够容纳多人活动；重新设计建筑物并增加它的功用，使其全年都能为市民提供聚会空间。

主要的建筑理念是给原有的建筑增加一座礼堂，使之能够在保存历史记忆的同时也能被广泛应用于各类公众活动，如演讲、电影放映、青年团体活动等等。

礼堂

礼堂通过其倾斜的立面将内部的几何结构展示于周围的公园和城市。这种几何形式设计的另外一层寓意还在于创造了一种具有象征意义的几何结构，它与保存记忆的概念以及这座建筑作为一座纪念碑的功能相联系。礼堂边缘巨大的窗户面向北方，从而将柔和的光线引入礼堂，并能让人观赏到激动人心的海法湾的美景。由一个富有诗意的寓意联系现实，窗户成了内部空间的一个明亮的结尾，借此来平衡丧失亲友的悲痛与鲜活世界的光明与希望。

缅怀空间

每位阵亡将士都有一个盒子，上面贴着他们的照片，悬挂在建筑内的缅怀空间的墙上。在盒子内，有一个地方是家人和政府用来存放有纪念意义的私人物品的。在这个缅怀空间里，本身就有一个地方可以坐下来与他人交流阵亡将士纪念册的内容以及对于他们的记忆。

缅怀空间的中心被狭窄的长窗照亮，这里原是用于战争防御位置的狭长的射击口。缅怀空间被设计为模块化的形式，因此，如果有需要的话，还可以另外再增加盒子。

装饰材料

与扩建部分相对的，原有建筑的轮廓则通过使用沉积在砂浆中的铝条而得到了凸显，与此同时也在新建部分中强调了旧建筑物的轮廓。本建筑的用料很简单。缅怀空间中的地面使用的是混凝土地板。整座建筑内的天花板均以橡木板覆盖，营造了一种温暖的氛围。礼堂区域也布满了木质的、带有海绵垫的坐椅。还为残障人士准备了一条专用通道。

广场

建筑物前方用于举行聚会的广场是在原有广场的基础上设计建造的。改建工作内容包括为残障人士的使用增加舒适度和便捷性，在边缘建造看台台阶以便提供更舒适的观赏位置，以及在广场的南端增加另外一个观景位置。广场上的饰面材料是一种显眼的混凝土，界定了台阶边缘，以使其区别于灰色的混凝土砖。

Nesher Memorial

Nesher Memorial was built on a basis of an historic preservation building that was used in the past as a guarding position. In "48 war the position was used as a protection to the residents of Giv'at Nesher. The building is located in the heart of a quiet neighborhood, on a hilltop, surrounded by a large square and a children's playground.

Prior to the construction works, the building was neglected and was used in favor of the public only once a year – in the National Memorial Day, then it was open for a tour and impression of memorial pictures, as a part of the public ceremony for the memory of the fallen.

Nesher Municipality asked SO Architecture office to expand the square surrounding the building so it will enable multiplayer events, and to redesign the building and add functions to it, so that it can serve the public throughout the year as a gathering space.

内谢尔战争纪念馆
SO Architecture

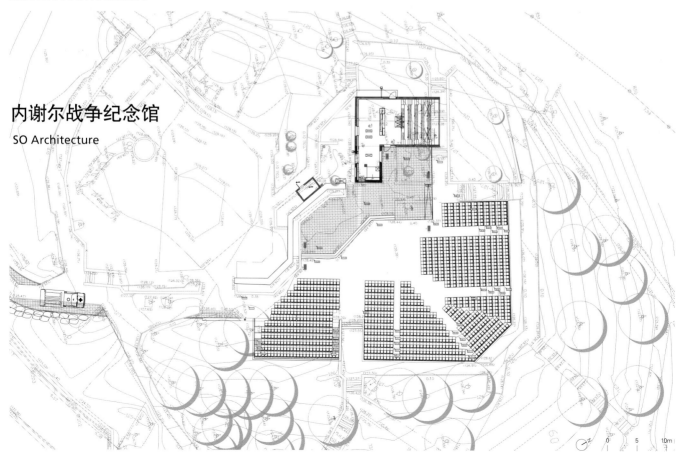

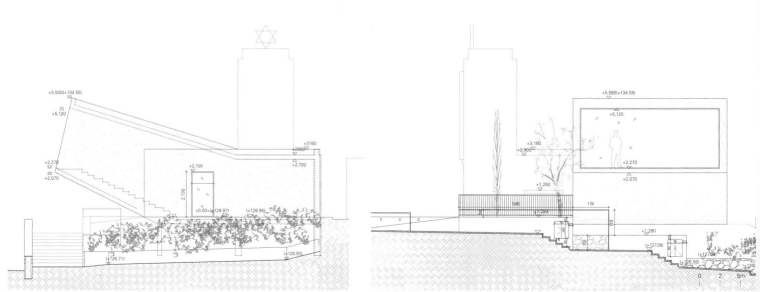

西北立面 north-west elevation 　　　　　　　　东北立面 north-east elevation

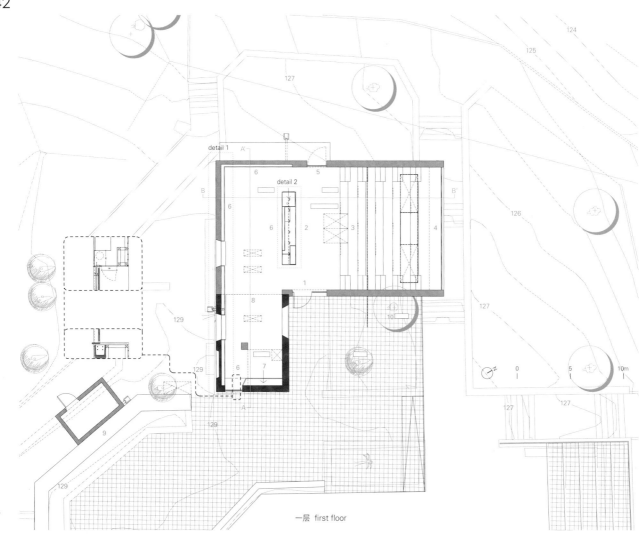

一层 first floor

1 建筑主入口
2 中央空调系统储藏室+其他电力元件
3 礼堂座位
4 临北窗眺望海法湾
5 紧急出口
6 阵亡将士肖像展示
7 纪念册
8 从旧建筑到新建筑的分界线
9 原先的户外纪念墙
10 原先的树木

1. main entrance to the building
2. center HAVC system storage + other electricity elements
3. auditorium seats
4. northern window view to Haifa Bay
5. emergency exit
6. display of the fallen portraits
7. memorial book
8. line showing the transformation from old to new building
9. existing outdoor memorial wall
10. existing tree

The main architectural idea was to add to the existing building an auditorium structure that could be used for different kinds of public activities such as lectures, film screenings, activities of youth groups, etc., along with preservation of the historical memory.

The Auditorium

The auditorium reveals its inside's geometry to the park and the city surrounding, by its inclined facade. This geometrical act has an additional meaning in creating a symbolic geometry that communicates with the memorial concept, and the function of the building as a monument. A large window is located at the edge of the auditorium, facing north and thus brings a soft light into the auditorium and enables a breathtaking view at the landscape of Haifa Bay. In a poetic allegory to reality, the window functions as a bright ending to the inner space, and thus symbolizes the balance between the bereavement pain and the light and hope in the living world.

The Memory Space

A box for each fallen, with its picture on it, is hanged on the wall of the memory area inside the building. Inside it, there is a room for storage of memorial personal belongings that the family and the municipality can put. In the space itself there is a place for seating and communion with the memory books and the memory of the fallen.

The center of the memory space is lighted by long and narrow windows that were the shooting slits in the original guarding position, The memory space was designed in a modular manner, so that if necessary, it will be possible to add additional boxes without any difficulty.

Finishing Materials

The original building's contours regarding to the addition, are marked and highlighted through aluminum bars that were sediment in mortar and emphasize the contours of the old building regarding to the new addition. The materials that we used in the building are simple. The floor in the memory space is a concrete floor. The ceiling along the whole building is covered with oak planks, so that it creates a warm atmosphere. The auditorium area was also covered with wood, and sponge padded seats. An access to disabled was also arranged.

The Square

The gatherings' square on the front of the building, was designed on the basis of the existing square. The works in it included suitability and accessibility to the disabled, creating grandstand steps in the edges to enable a more comfortable viewing, and an addition of another viewing site at the southern end of the square. The finishing material of the square is a visible concrete that delimits the steps, and gray concrete blocks.

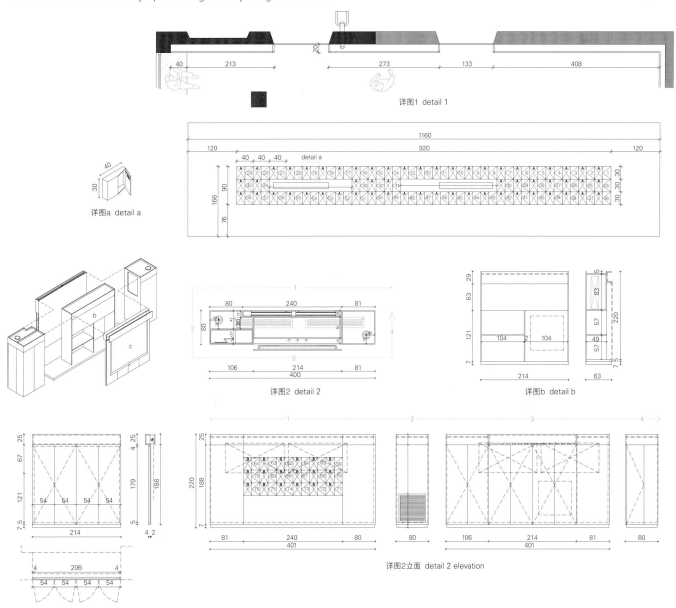

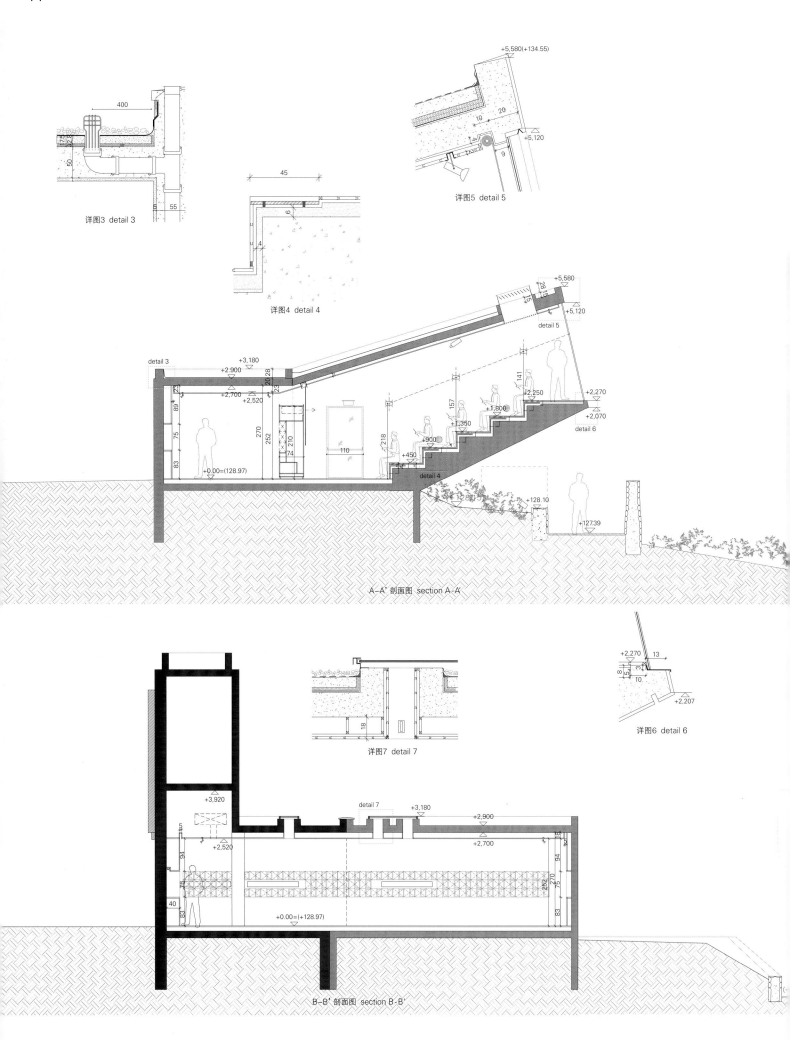

项目名称：Nesher Memorial
地点：Nesher, Israel
建筑师：SO Architecture
项目团队：Shachar Lulav, Oded Rozenkier,
Alejandro Fajnerman, Tomer Nahshon, Samer Hakim
用地面积：1,100m²
建筑面积：118m²
竣工时间：2013
摄影师：©Shai Epstein (courtesy of the architect)

社区的起源：独立个体

The Source of Com Individual Bodies

社区是什么？我们也可以这样问，说到社区，我们赋予社区怎样的意义？问题的答案将涉及一个概念，因此也会涉及到思想。如果我们身处社区之中，那么我们的身体就会告诉我们社区是什么。无需去思考，我们就会知道我们属于社区的一部分，或者不属于。

没有个体就不能称其为社区，而每个个体都有自己的身份，也都有自己的规格。个体的身体是一种规格，一小群人是另一种规格，而更大一些的人群就可能构成一个社区。

在整个建筑史中，可以说直到我们所在的时代，我们都建造属于社区的建筑，这样说是因为这些建筑物第一次为生活其中的个体所考虑。

拥有身体的感觉是什么样子？有了这方面的知识，我们就为我们的身体创造空

What is community? We could ask what we mean by it, but the answer would involve a concept, and therefore thought. If we experience community, then we know with our body what it is. We don't think about it. We know we're part of a community – or not.

We can't have a community without individuals, and individuals have their own identities. Individuals have their own sizes too. The body is one size; a small group of people is another size; and a larger group may become a community. Throughout our experience of building – until perhaps our own current age – we've built buildings that make place for communities because they first make place for individuals.

We were in touch with our knowledge of what feels like to have a body. We made spaces for our bodies. And our architecture proves it.

艾琳娜·加罗文化中心_Elena Garro Cultural Center/Fernanda Canales + Arquitectura 911sc
隆勒索涅城市媒体中心_Lons-le-Saunier Mediatheque/du Besset-Lyon Architectes
Daoíz y Velarde文化中心_Daoíz y Velarde Cultural Center/Rafael De La-Hoz Arquitectos
Fjelstervang户外社区中心_Fjelstervang Outdoor Community Hub/Spektrum Arkitekter
友好中心_Friendship Center/Kashef Mahboob Chowdhury/Urbana
新Encants市场_New Encants Market/b720 Arquitectos

社区的起源：独立个体_The Source of Community: Individual Bodies/Jaap Dawson

间，而我们的建筑证明我们的感觉。

　　从一根柱子中，我们看到的是我们自己的身体。从一行柱子中，我们看到的是整个社区。社区包含的不仅仅是大量的柱子，还包括柱子之间的空间以及柱子的规模。在这样的空间中，我们的身体或者感觉非常舒服，或者没有亲近感。

　　如果我们没有在一栋建筑物中的身体体验，那么凭着我们对建筑元素的认同，我们可以查阅别人的文献，来看看他们怎样体验建筑，如何把建筑元素类比为我们的身体的，这些文献包括约瑟夫·里克沃特（《跳舞的柱子》）、杰弗里·斯科特（《人文主义建筑》）、乔治·赫西（《古典建筑意义的缺失》）。

　　但要想重新发现我们如何体验柱子、空间、建筑，最好的方式就是简单地居住其中。而为了亲身体验，我们需要能够把我们和我们的体验重新连为一体的建筑。

In a column we see and meet our own body. In a row of columns we see and meet a community of bodies. It's not only a number of columns that constitute the community: it's the space between them, the scale of the measures as well. Either our body fits comfortably in the space, or else it feels alienated.
If we've lost touch with our bodily experience in a building, with our identification with building elements, we can always consult what other people have written about experiencing a building element as an analogy for our body: Joseph Rykwert (*The Dancing Column*), Geoffrey Scott (*The Architecture of Humanism*), George Hersey (*The Lost Meaning of Classical Architecture*).
But the best way to rediscover how we experience a column, a space, a building with our body is simpy to live it. And in order to live it, we need buildings that reconnect us with our experience.

意大利的坎皮多里奥广场，米开朗基罗·博那罗蒂设计，1547年
Piazza del Campidoglio, Italy by Michelangelo Buonarroti, 1547

荷兰的圣本笃修道院，道姆·凡·德·兰设计，1967年
St. Benedict's Abbey, Netherlands, Dom Van Der Laan, 1967

坎皮多里奥广场

米开朗基罗设计的坎皮多里奥广场如同一位伟大的老师。窗户两侧的柱子和我们的身体一样高。还有这一开阔门窗结构两侧的柱子：让我们感到符合人体尺度的社区在其间的和谐。最后就是巨大的壁柱。因为我们已经感受到——因此已经知道——其他柱子之间的空间的大小、整个广场社区的其他构成，因此我们现在能够感觉并认识整座建筑的规模。在广场中，人如同可移动的柱子，几乎与建筑物的柱子融为一体，无法区分。

圣本笃修道院

荷兰的圣本笃修道院教堂（1967）位于Vaals附近，体现了米开朗基罗的专业知识。修士设计师道姆·凡·德·兰从厚重的建筑元素的使用和空间两个方面向我们展示了社区的意义。凡·德·兰认为，如果空间的宽度由一定量的柱子或墙壁厚度构成，我们就可以感觉到并知道空间的大小。柱子和墙壁都有厚度，就像我们的身体一样。如果我们能感觉和感知到居住空间里由个体组成的社区的存在，那么，我们就会感到此处空间很亲切，就会感觉到身体与空间之间的联系。

一旦我们了解了符合人体尺度的柱子之间的空间宽度，我们就可以构建更大一点儿的空间——大一点儿的社区，但仍然保留与最初的柱子，即原来的身体的关系。更大的空间与原始的建筑体块是有联系的。空间个体构成的可识别的社区又将成为新的更大一些空间内社区中的个体。毕竟，只有我们见过的并定义过的空间之间的空间才富有活力。

卡德里亚诺

卡德里亚诺的住宅社区位于博洛尼亚市郊，向我们展示了如何能够以适度而历久弥新的方式来建造个体和社区空间。马西迪工作室将我们带回到2009年，让我们亲身体验一下这一社区的设计。犹如个体的柱子形成一面多孔墙，成为一个社区。显而易见，这面墙的后面和上面的房屋是较大规模的个体：共同形成一个社区。我们构建的建筑没有必要太抽象，太概念化，没有必要故意让人不熟悉。

上述三个例子把我们与我们的体验重新连为一体。我们清醒地或半清醒地认识到，我们可以从我们的建筑中看到我们自己的影子。我们也知道个体、社区的一员是什么概念。其设计作品得到专业新闻媒体关注的其他建筑师们是否也知道这一点呢？

你自己做判断。不要想，不要寻找概念。直接去接触空间，用你的身体接触空间，体验那些空间的边界，即定义空间的那些厚重元素，进

Campidoglio

Michelangelo's Campidoglio is a grand teacher. The columns flanking the windows have the scale of our body. Then there are the columns flanking the open bays: they help us feel the community of human body sizes that fit between them. And finally there are the giant pilasters. Because we've already felt – and therefore known – the sizes of the spaces between the other columns, the other members of the community, we're now able to feel and know the scale of the whole building. The building's columns are nearly indistinguishable from the people as living columns in the square.

St. Benedict's Abbey

The Abbey Church of St. Benedict (1967) near Vaals, the Netherlands, embodies Michelangelo's expertise. Dom Van der Laan, the monk-architect, shows us what a community is both in terms of massive elements and in terms of space. We can feel and know the size of a space, Van der Laan discovered, if the width of the space is composed of a number of columns or wall thicknesses. Columns and walls have thickness, just as our body does. If we can feel and perceive a community of bodies in the width of the space we inhabit, then we have an intimate and bodily relationship with that space.

Once we've established the knowable width of a space between bodily columns, we can build larger spaces – larger communities – that retain their relationship with the original column, the original body. The larger space has a relationship with the first spatial building block. That recognizable community of spatial bodies becomes an individual in the new community of the larger space. That space, after all, comes to life between the defined spaces we're already met.

Cadriano

The community of houses in Cadriano, just outside of Bologna, shows us how we can build both individuals and communities in a modest and timeless way. The Studio di Via Masi brings us in 2009 back to our bodily experience. Columns as individual bodies can become a perforated wall that's now a community. And the individual houses behind and above that wall are clearly individual bodies of a larger scale: in consort they form a community. What we build need not be abstract, need not be conceptual, need not be expressly unfamiliar.

These three examples reconnect us with our experience. We know, consciously or half-consciously, that we see ourselves reflected in what we build. We know too what it is to be an indi-

意大利博洛尼亚市的卡德里亚诺的住宅社区
community of houses in Cadriano, Bologna, Italy

而发现建筑会在多大程度上体现构成社区的个体。只有到那时,你才能反思建筑对人类社区意味着什么。

艾琳娜·加罗文化中心

走进文化中心,首先映入眼帘的是巨大的玻璃幕墙。墙上没有任何有厚度的元素。我们的身体与墙不产生任何交集。

一旦进入其内,我们看到了原始的房子,但是房子和玻璃幕墙之间的居住空间没有任何元素能告诉我们房子的大小,也没有任何个体提示我们,它们共同形成一个社区。

书架依旧是书架,只有两种尺寸、两种规模:每本书的大小和整排书架墙的大小。没有坎皮多里奥广场中那样的中等规模,我们感觉不到每本书与整排书架墙之间的联系。

进入玻璃幕墙,我们位于原始房子的外面。房子的门廊迎接着我们,让我们知道房子的原始立面有多大,但身居在这栋新建筑物之内,却感到置身其外,我们的身体感觉不到,也认识不到位于某个空间之中。

Daoíz y Velarde文化中心

在建筑的外面,我们与社区对视:形成立面的模块拥有我们所知的符合人体尺度的房间。一个人如果位于一间规模符合个体居住的社区的房间内时,便会有如归的感觉。

然而,如果我们走进新建筑的内部,我们仅仅是看见一些事物。我们看见的是一些机械构件,而不是重新组装成个体的元素。所有的新柱子和横梁都保留了垂直或水平的桁架。当然,它们都有一定的厚度,但是它们并没有邀请我们去辨别,更不用说环抱它们。如果这里没有我们可以认为是个体的元素,我们也几乎无法去体验一系列的、可能形成社区的个体。

Fjelstervang户外社区中心

"建起来!"这是一座丹麦小村庄内的村民的座右铭。这些村民帮助建筑师建成了一座全新的村务大厅。这些人形成了一个社区,而建筑师设计的这座建筑则反映了这个社区。具有可持续性的木构件符合聚在一起工作的人们的人体尺度,来形成适合一群人的空间。我们对将部件结合在一起的方式有所了解,同时也对用双手建起这座建筑的人们有所理解。

这一建筑综合体不仅仅是各个部分构成的社区,也是小型建筑组成的社区。它们组合到一起,来形成一个整体:一个由小型社区组成的

vidual and what it is to be a member of a community. Do other architects, whose buildings have received attention in the professional press, know that too?
Be the judge yourself. Don't think. Don't look for a concept. Just meet the spaces. Meet the spaces with your body. Meet the boundaries of those spaces – the massive elements that define them. Discover to what extent the building embodies individuals who can comprise a community. Only then will you be able to reflect on what the building means for the community that is humankind.

Elena Garro Cultural Center
When we approach the building, we meet an expansive glass wall. The wall has no elements with thickness. Our body doesn't meet any parts of the wall as other bodies.
Once inside we meet the original house. But the space we inhabit between the house and the glass wall has no elements that tell us how big it is. It has no bodily individuals that tell us they form a community.
The bookshelves remain bookshelves. They give us two sizes, two scales: the size of the individual book and the size of the whole bookshelf wall. There's no intermediate scale as there is in the Campidoglio. We can't feel the relationship between the individual book and the whole wall of shelves.
When we're inside the glass wall, we're outside the original house. The original house has a porch that greets us as a body. It gives scale to the original facade. But inside the new building we feel we're outside, not contained in a space our body can feel or know.

Daoíz y Velarde Cultural Center
On the outside of the building we come face to face with community: the modules that form the facade have the size of rooms we know with our bodies. An individual body feels at home in a room whose size is a community of individual bodies.
When we go inside the new building, however, we meet only things. We meet machine parts rather than elements we can recognize as bodies. All the new columns and beams remain vertical or horizontal trusses. They have a certain thickness, to be sure, but they don't invite us to identify with them, let alone hug them. If there are no elements we can greet as individual bodies, we hardly can experience a collection of bodies that might form a community.

Fjelstervang Outdoor Community Hub
"Build it up!" was the motto of the people in a small Danish village

从规模上来说，友好中心的独立亭子和我们的身体一样大小。
Individual pavilions of Friendship Center like our own body in terms of scale

社区。这个村落大厅包含一个社区，拥有清晰定义的空间。这是一个主要的建筑典范，不仅仅对于其他建筑师来说，而且对于使用建筑、辨别建筑、在建筑内享受的人们来说同样如此。

友好中心

友好中心建筑将我们与我们传递友情的能力连在一起：与另一个人的友情，与另一群人的友情，与另一个社区的人的友情。支撑独立亭子的柱子向我们展示了它们所承受的压力。这些独立的亭子形成了第一社区，其间带有一些可辨认的空间。之后这些独立的亭子和房间连接着扶手，形成一个更大的社区：一座建筑。最后，这组建筑组合在一起，发挥功能，或者交织在一起，形成一个综合体，我们将其定义为村落。

本地的砌砖对社区进行了加固，因此我们无需在长距离的材料运输中消耗精力，砌砖既没有喧宾夺主，也显得十分优雅。它们擅长在重压之下挑起重担，同时也提醒我们是这些个体，聚在一起成为一组施工者，将砌砖垒砌起来，形成这些空间。丰富且影响力较大的文化教会我们将这个世界分为发达区域和发展中区域，而友好中心产生于所谓的发展中区域。但是其结构、材料以及施工者却远比我们所熟悉的一些建筑单位更胜一筹。友好中心是一座永恒且全新的建筑，如同我们之间的友谊一样，我们从中获益良多。

隆勒索涅城市媒体中心

无需思考，我们便迈向了这座建筑。教堂横穿其中，我们即刻便有如家的感觉。为什么呢？因为我们的身体感受到了拱壁作为一个个体，成为我们所看见的空间的一部分，也因为我们感受建筑，如同一个充满活力的个体感受一个有机成长的身体一样。

之后我们才注意到新建筑。在室外，我们看见一堵十分抽象的墙，带有抽象的、精确的孔洞，我们是正在看一个来自于20世纪60年代的计算机存储卡吗？

而在室内，我们感受到了墙壁的厚度。但是那些墙体上的洞口和我们有什么关系呢？我们的身体陷入了一张网内，我们在其中无法看见其他身体，而我们的身体也被缚住。这种体验便是这座建筑的建造目的吗？

新Encants市场

在温暖的天气里，处于一个大型市场内，还有什么比头顶上的屋顶

who helped build their new village hall. The individuals formed a community, and the building the architect designed reflects that community. The sustainable wooden elements all have the scale of individuals working together to form spaces for groups of individuals. The means of joining the parts together are visible and understandable for people building with their hands.

The complex is not only a community of parts but a community of small buildings as well. They all work together to form the whole: a community of smaller communities. The village hall embodies community in its clearly defined spaces. It is a prime example of architecture not chiefly for other architects but for the people who use, enjoy, and identify with the building.

Friendship Center

The Friendship Center puts us in touch with our capacity to give and receive friendship: friendship with another invidivual, with a group of individuals, with a whole community of groups. The columns that form the individual pavilions show us the loads they bear. The individual pavilions form the first community of bodies with knowable spaces between them. Then the individual pavilions and rooms link arms, forming a larger community: a building. Finally, the clusters of buildings work together – or rather, they play together – to form a complex we recognize as a village. Locally made bricks reinforce the community. We don't need to waste energy in transporting complex materials over long distances. The bricks do their work modestly but elegantly. They excel at carrying their loads under compression. They remind us too that individual bodies, working together as a community of builders, laid the bricks that form the spaces.

Rich and powerful cultures have taught us to divide the world between developed and developing areas. The Friendship Center has grown literally out of the ground of the so-called developing world. But its structure, its materials, and its builders may well prove far more developed than the architecture practice we're familiar with. The Friendship Center is as timeless and new as every new friendship we form. We can learn a great deal from it.

Lons-le-Saunier Mediatheque

Without thinking, we walk toward the building. Our body feels immediately at home with the church across from it. Why? Because it senses the buttresses as bodies that form segments of space we know at once. Because it meets the building as a living body meets a body that grew organically.

Then we notice the new building. On the outside we see abstract

Fjelstervang户外社区中心，运用符合人体规模的可持续木构件建造而成
Fjelstervang Outdoor Community Hub, structured with sustainable wooden elements of individual scale

更加令人期待的吗？前所未有的绝佳屋顶应该是什么样子，是从远处看便能吸引人们的眼球的吗？

当我们去市场的时候，我们不仅仅能看见一些货物，还会看见许多其他人们：张望的人们、谈话的人们、卖货的人们以及买货的人们。市场是一个社区，每个人为了一个目的聚到一起，但是因为有许多人聚在一起，因此便会产生更多的目的、更多的体验，而不是仅仅的买和卖。

那么我们都看向哪里？在我们的左侧和右侧，我们看见人们正在卖的货物，我们还能看见那些希望我们掏钱的人们。充满活力的人们和货摊为我们穿行的地方赋予了规模，坦白地说，我们几乎望不见屋顶。

事实上，我们在什么时候望向屋顶，或者是天花板？如果天花板成拱状，且带有肋条，以提醒我们也有自己的肋骨。如果是方格天花板，则让我们的身体感受到它是怎样蔓延到墙壁（界定且形成了我们所处的空间）上的。亦或者，如果我们位于Cistine教堂，便实在是没有可看的地方了。

新Encants市场是不同寻常的，但是它并没有与人体相接触，而人们也没有认为它是不同寻常的。它仍然保持着一个装饰过的帆布的形象，延伸在社区内彼此为生活奔波的人们的头顶上。

建筑内的社区和个体

当我们注意到一座建筑时，当我们居住在空间内时，我们的身体有话语权。我们应该设计什么，并不是我们的理念；我们要证明我们的设计是现代的，也并不是我们的意愿，而是我们的生活、我们的身体，才是我们的理念和意愿。我们的身体包含我们的灵魂，也包含我们内在的感知和了解方式。

如果我们是个体，我们便会形成一个社区，但我们必须是带有身体和灵魂的社区。事实便是如此。如果我们建造的世界是要反映我们是谁，那么它不应该赋予我们身体和空间，以提醒我们也拥有个人的身体和群体的身体吗？

walls with abstract, mathematical holes in them. Are we meeting a computer card from the 1960s?
On the inside we sense the thickness of the walls. But what do the openings in those walls do with us? Our body feels enmeshed in a net. We can't meet a body in it. And our own body feels constrained. Is this experience the goal of our architecture?

New Encants Market

What more could you wish for than a roof over your head in an expansive market in a warm climate? And what better roof could there be than a roof you'd never seen before, a roof designed to attract attention from afar?
When we go to a market we meet not only the wares but a host of other bodies: bodies looking, bodies talking, bodies selling, bodies buying. A market is a community of individual people coming together for a single purpose; but because there are so many people who've come together, there are many more purposes, many more experiences, than only selling and buying.
Where do we look? To our left and to our right, at the things people are selling and at the people who want us to part with our money. The living bodies and the market stalls give measure to the world we walk through. If we are honest, we rarely look up at the roof. When do we in fact look up at a roof, or rather a ceiling? If the ceiling is vaulted, and therefore has ribs that remind us of our own ribs. If the ceiling is coffered, letting us feel with our bodies how it spans the walls that define and make the space we're in. Or if we're in the Cistine Chapel – but there's virtually no other place to look.
The Els Encants Vells is breathtaking, but it doesn't touch real bodies who need to take breaths. It remains a decorated canvas, thankfully stretched over the heads of people busy with their lives in community with each other.

Community and Individual in Architecture

When we regard a building, when we live in a space, it's our body that has the last word. It's not our notion of what we should design. It's not our wish to prove we're contemporary. It's our life. It's our body. And our body contains our soul, our inborn manner of perceiving and knowing.
We can form a community only if we're individuals: individuals with bodies and souls. That fact is given. And if the world we build is to reflect who we are, shouldn't it give us bodies and spaces that remind us of our own body and groups of bodies? Jaap Dawson

都市与社区——社区的起源：独立个体 Community and the City—The Source of Community: Individual Bodies

艾琳娜·加罗文化中心

Fernanda Canales + Arquitectura 911sc

　　艾琳娜·加罗文化中心坐落在墨西哥城南部科约阿坎城镇的历史街区，是为了缅怀这位墨西哥作家而建造的。

　　本项目的任务是保存一座已列为文化遗产的20世纪早期的建筑，并将其改建成多功能活动中心，因此需要增加空间并将原建筑的功能重新整合，使它容纳现代化的时尚书店、礼堂、餐馆、会议室以及阅读空间。无论建筑内部还是外部都被列为新文化中心功能空间的一部分。

　　工程保留了既有的独立式房屋，并完善了房屋内部及前后的两处扩建部分的功能，旨在保存并提升原建筑的特色。

　　房屋正面临街，扩建部分使用混凝土框架镶嵌了玻璃幕墙，用作公共休息厅，重新建立起建筑物和街道之间的联系。

　　双层高的墙壁被用作了书架，从人行步道看过来形成了一个斜向的视角，提醒了人们文化中心的存在。从二层延伸出来的两个平台便于市民由此接近新的混凝土墙壁上的书架。

　　拆除了一面将建筑物与街道分隔开的墙壁，并用一个新花园充当了人行步道的延伸。在这个屋前的扩建部分，原有的树木被保留了下来。在屋顶安装了一些天窗，让这些树木可以继续生长，同时为屋内采光。

　　在原有房屋的内部增加了一系列孔洞，创造了双层高的空间和阅览室。一段木质台阶通向二楼，其规模和材质使其成了建筑内的一个重要组成。在原有建筑与扩建部分相连的入口处，钢板框架与厚重的砖石墙体形成了鲜明的对比。

　　基地后方的扩建部分是一个现场浇筑的混凝土地下停车场，而会议室、办公室、卫生间和小礼堂则分布在两个主楼层。办公室和会议室面向建筑西南角的一个庭院开放，这为它们提供了自然采光。在建筑北部有一座用加勒比胡桃木装饰的小礼堂，里面配备有折叠坐椅，方便多种布局和活动。

　　建筑物北侧有一个可以在此读书的花园，这里植被茂盛，地面铺着

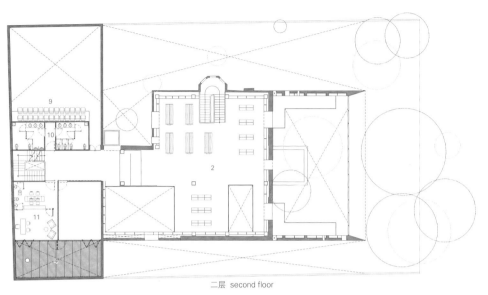

二层 second floor

1 入口 2 图书馆 3 阅读室/游乐场 4 露台 5 教室 6 多功能室
7 花园 8 户外论坛 9 夹层 10 卫生间 11 办公室
1. access 2. library 3. reading room/playground 4. terrace 5. classroom
6. multipurpose room 7. garden 8. outdoor forum 9. mezzanine 10. bathroom 11. office

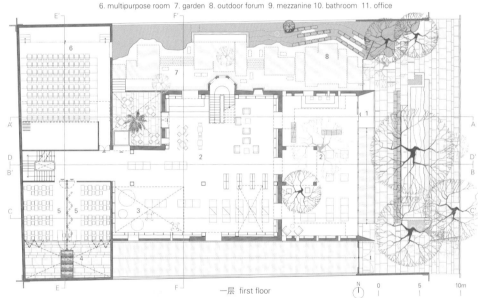

一层 first floor

Entorno Taller de Diseño设计的软软的面砖，还有一对巨大的青铜浇铸的大门；外部空间通过餐馆与原建筑相连，这里还有由Paloma Torres设计的艺术品。

有限的材料色彩、零散的体量切入、新旧建筑以及与外界之间的衔接都在试图赋予这座古老的建筑以崭新的生命，在为小区创造了一座地方文化工程的同时也为艾琳娜·加罗的遗产增色。

Elena Garro Cultural Center

Elena Garro Cultural Center is located in the historical district of Coyoacan, the southern part of Mexico City honoring the memory of the Mexican author.

Faced with the task of preserving a heritage-listed house of the beginning of the 20th century and transforming it into a multi-activity cultural center, the project deals with adding spaces and re-functionalizing the existing building in a contemporary fashion bookstore, an auditorium, cafeteria, seminar rooms as well as reading spaces, both interior and exterior are part of the program for the new cultural center.

The detached house is preserved and re-functionalized from the inside as well as two additions are built to the front and to the back in a strategy that strives to preserve and heighten the character of the existing building.

Towards the front of the street, an extension is built as a concrete frame with a glass facade that serves as a public foyer and re-establishes a relationship between the building and the street. The double-height walls are used as bookcases creating diagonal views from the sidewalks that allude to the program of the cultural center. Two platforms extend from the second level of

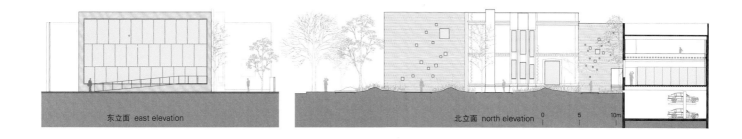

东立面 east elevation　　　　北立面 north elevation　0　5　10m

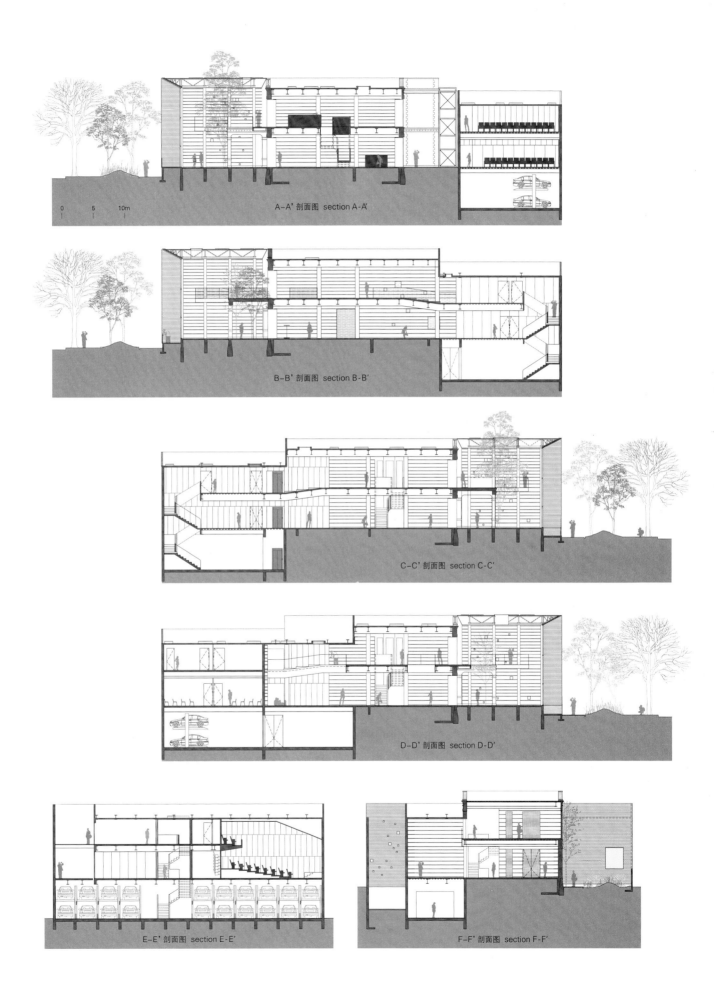

the house to give public access to the bookshelves on those new concrete walls.

The demolition of a wall separated street from property and a new garden acted as an expansion of the sidewalk. Inside this front extension, the existing trees are preserved and a series of skylights in the roof allow them to keep growing and simultaneously bring in light.

The interior of the existing house is transformed with a series of perforations that create double-height spaces and reading rooms. A wooden-clad stair leads to the second floor and through its scale and materiality becomes a relevant piece within the house. The openings where the existing house connects to the new additions, are framed with steel plates that contrast with the thick masonry and brick walls.

The addition at the back of the site is a cast-in-place concrete volume that houses parking in the basement and seminar rooms, offices, restrooms and a small auditorium in the two main floors. Offices and seminar rooms face a courtyard on the southwest corner of the building bringing in natural light. On the northern section, a small auditorium cladded in tzalam wood and with a foldable seating allows for multiple configurations and events.

On the northern side of the property is a reading garden with lush vegetation and soft paving designed by Entorno Taller de Diseño and a large cast-in-bronze door; an artwork by Paloma Torres that configures an exterior space that connects to the existing building through the cafeteria

The limited material palette, discrete volumetric interventions, the articulations between old and new and the exterior spaces attempt to give a new life to the existing property, creating a local cultural program for the neighborhood and honoring the legacy of Elena Garro.

项目名称：Elena Garro Cultural Center
地点：Mexico, Mexico City, Fernández Leal no. 43, Col. Barrio de la Concepción, Delegación Coyoacán, C.P. 04020
项目设计：Fernanda Canales, Arquitectura 911sc
设计负责人：Fernanda Canales, Saidee Springall del Villar, Jose Castillo Oléa
设计团队：Iván Cervantes, Edgar Romero, Arturo Carreón, Anabel Chavez, Javier Juárez
结构：Grupo SAI
服务系统：Eng. Carlos Medina
照明：Artenluz / 景观：Entorno Taller de Paisaje
艺术装置："The Forest Transformed": Paloma Torres, (Bronze doors), Book Lamps by Ariel Rojo
甲方：Educal / Conaculta
建筑面积：1,358m²
材料：concrete walls, exposed wooden formwork finish, steel columns and mezzanines, tempered glass, tzalam wood wall paneling and flooring, acid-polished granite flooring, door and window frames plated black steel with automotive paint, lacquered mdf bookcases, bronze sculpture, outdoor plazas in laminated stone, wooden sleepers and various types of gravel combined with existing and new vegetation appropriate for the context
竣工时间：2012.10
摄影师：
©Jaime Navarro(courtesy of the architect)-p.52~53, p.57, p.60, p.62, p.63
©Sandra Pereznieto(courtesy of the architect)-p.54, p.56, p.58, p.61

现有建筑:由教堂、巴莱酒店和监狱围墙构成的这个市区环境,拥有简单的线条和布满矿石的墙体,气势宏伟,令人印象深刻。遗憾的是,从街道一侧无法体验这三座建筑之间的构建关系。

显然,在北面加入第四个元素能让这个城市的结构更完整。

缺少的元素正是图书馆和电影院,而新建筑优雅的曲线恰好为街道一侧留出了公共空间。在拥挤的市区环境中能创造出如此通风良好的设计,着实让人欣喜。

和周围建筑一样,新建筑的线条简洁明了,混凝土矿物质的使用和用量相得益彰。

与周边环境活跃的关系赋予了它非凡的力量和出众的特点。其南面弯曲的墙体和教堂顶部的巨大斜板遥相呼应;弯曲的设计自然地连接了可德利尔街道、新中心区域和教堂背面。

新建筑既尊重了周围的环境,也体现出自己的独特性。多个曲线的混合促成了美学效果,而整个建筑看起来又好像是一个自主的物体。站在它的任何一扇门前,都可以纵观到它的系统性,可以理解到这个方法在每个部分的应用,从而可以感知整个建筑。

室外的建筑表达同样体现在室内。从入口,人们就可以感受到室内的整体性。室内的布局清晰、大气,电影院的三层结构功能各不相同,还有一层地下室。

Lons-le-Saunier Mediatheque

The existing buildings: the church, the Hotel de Balay and the wall of the prison form an impressive urban setting made up of simple lines, great mineral surfaces and imposing masses. Unfortunately, the relation that these three constructions maintain cannot be experienced from the street.
Obviously, a fourth element is needed on the north side to complete the urban structure.
The missing elements are the library and cinema. The new building has the courtesy to curve itself in order to spare a public place that opens on the street. An urban respiration is created which is welcome in a very dense built context.

隆勒索涅城市媒体中心
du Besset-Lyon Architectes

西立面 west elevation

南立面 south elevation

Like its neighbors, the lines of the new building are simple, its matter is mineral(concrete) and its presence is stated without superfluous arrangements of volumes and materials.

It draws its strength and its strangeness owing to the fact that it maintains an active relationship with its neighbors. Thus, the curvature of its southern facade addresses a direct response to the large slope of the slate of covered roof of the church; the curve of its plan constitutes a natural connection between the street of Cordeliers, the new central place and the back of the church.

If it is respectful of its environment, the new building is also keen to assert its unique presence. Its aesthetic is regulated by a combination of curved lines. By this process it appears an autonomous object. Placed in front of one of his frontages, one seizes the system that regulates it, understands that it applies to the unit and perceives the whole.

The architectural expression of the exterior is found in the interior. From the entrance, one perceives the entirety of interior volume. The interior organization is of a great clearness: the various functions are distributed on three levels plus a basement for the movie theaters.

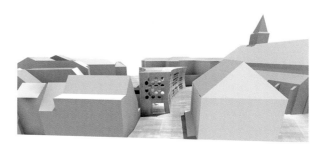

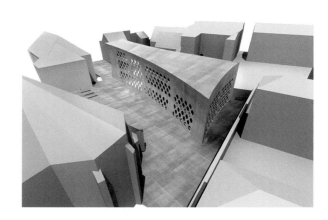

1 电影院（77座+3个残障人士座位）
2 放映室
3 电影院（182座+5个残障人士座位）
4 儿童厅
5 青少年厅
6 图书馆接待处
7 公共大厅

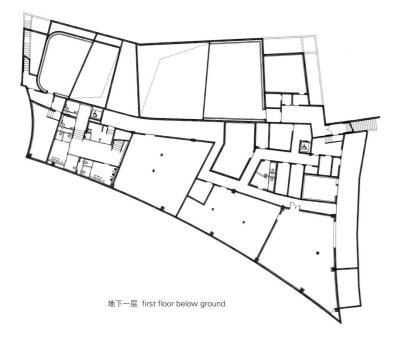

地下一层 first floor below ground

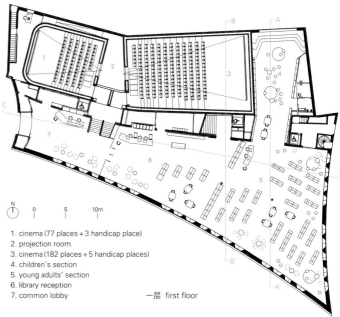

1. cinema (77 places + 3 handicap place)
2. projection room
3. cinema (182 places + 5 handicap places)
4. children's section
5. young adults' section
6. library reception
7. common lobby

一层 first floor

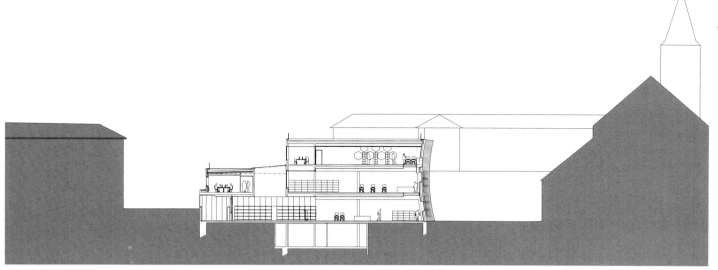

A-A' 剖面图 section A-A'

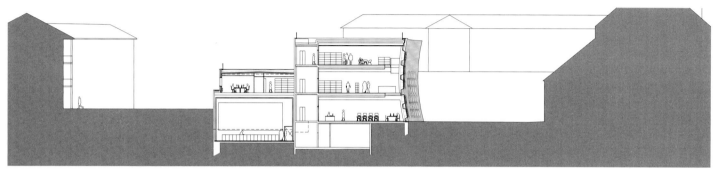

B-B' 剖面图 section B-B'

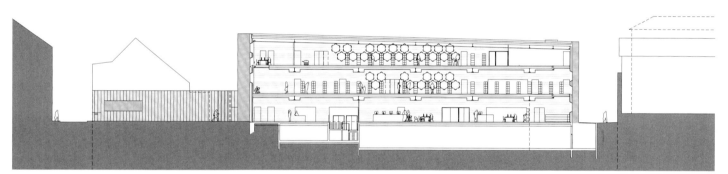

C-C' 剖面图 section C-C'

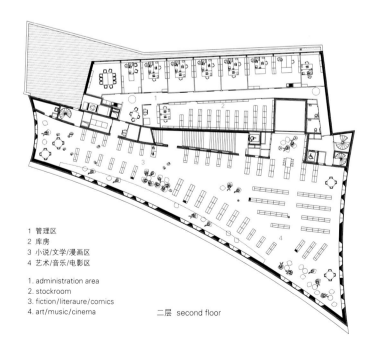

1 管理区
2 库房
3 小说/文学/漫画区
4 艺术/音乐/电影区

1. administration area
2. stockroom
3. fiction/literaure/comics
4. art/music/cinema

二层 second floor

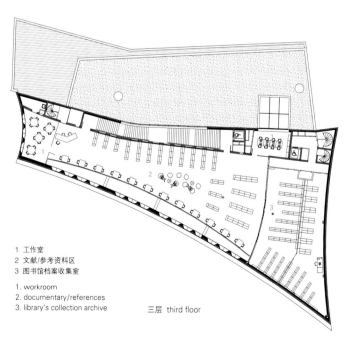

1 工作室
2 文献/参考资料区
3 图书馆档案收集室

1. workroom
2. documentary/references
3. library's collection archive

三层 third floor

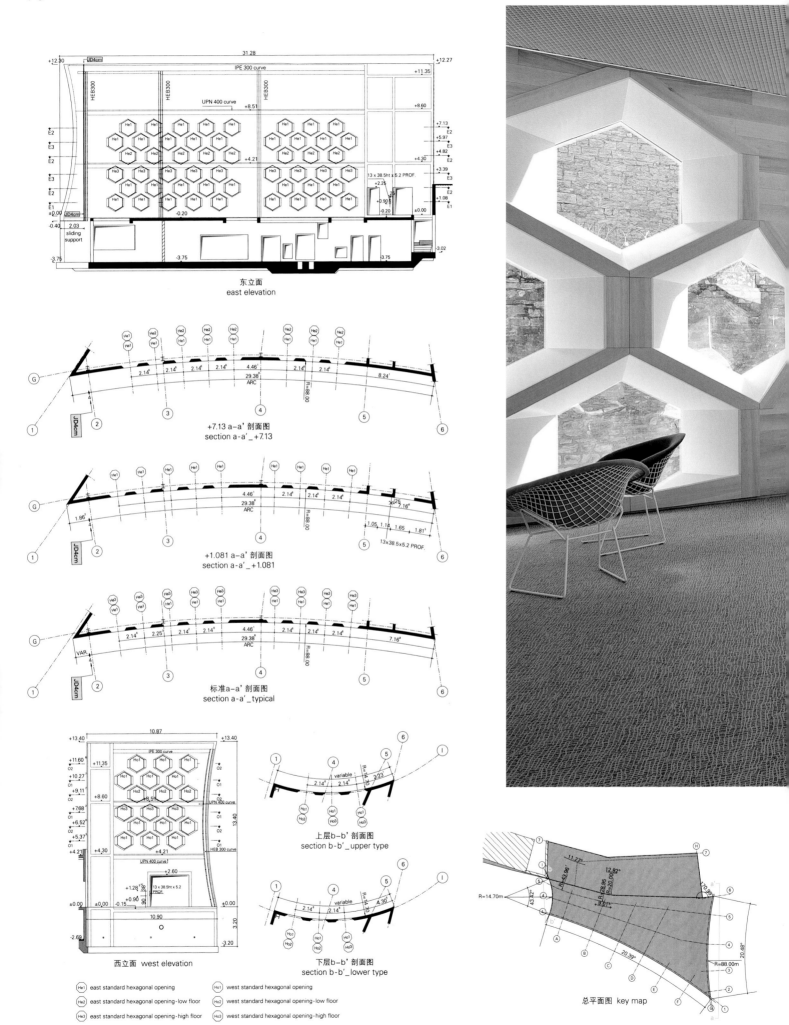

项目名称：Lons le Saunier
地点：rue des Cordeliers à Lons-le-Saunier, France
建筑师：du Besset-Lyon Architectes
设计团队：Dominique Lyon, Pierre du Besset
项目经理：Anne Tellier
结构工程师：KHEPHREN Ingénierie
流体工程：ESPACE TEMPS
经济学家：Jean-Claude DRAUART
环境质量顾问：Franck BOUTTE
执行建筑师：Véronique RATEL
甲方：Communauté de Communes du Bassin de Lons-le-Saunier
功能：library, cinema
面积：3,500m²
竣工时间：2012.11
摄影师：©Philippe Ruault(courtesy of the architect)

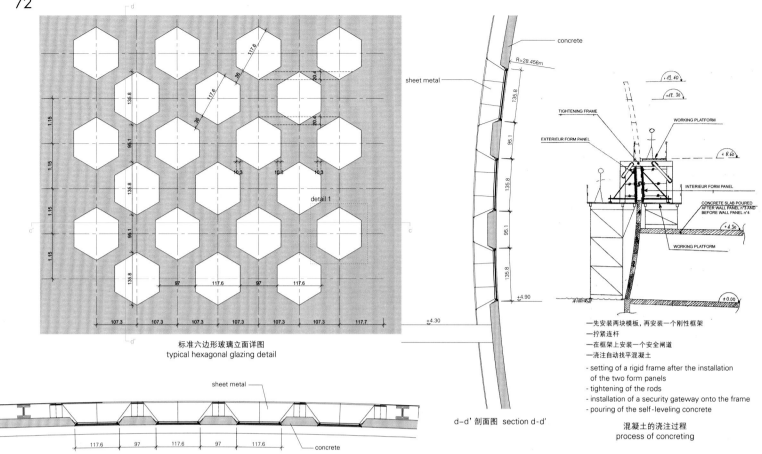

标准六边形玻璃立面详图
typical hexagonal glazing detail

c-c' 剖面图 section c-c'

d-d' 剖面图 section d-d'

— 先安装两块模板,再安装一个刚性框架
— 拧紧连杆
— 在框架上安装一个安全闸道
— 浇注自动找平混凝土

- setting of a rigid frame after the installation of the two form panels
- tightening of the rods
- installation of a security gateway onto the frame
- pouring of the self-leveling concrete

混凝土的浇注过程
process of concreting

详图1 detail 1

f-f' 剖面图 section f-f'

e-e' 剖面图 section e-e'

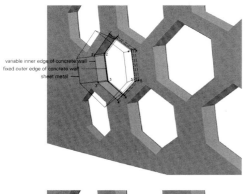

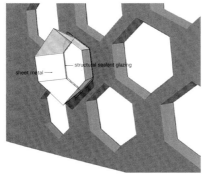

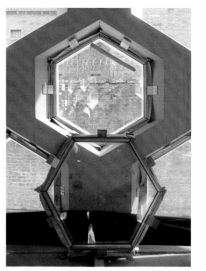

Daoíz y Velarde文化中心

Rafael De La-Hoz Arquitectos

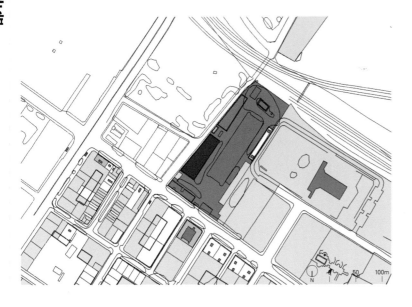

本项目是Daoíz y Velarde军营综合楼的一部分，是马德里工业和军事遗产的典型样本，所以首要目的就是保护建筑。

从一开始，设计理念就是要尊重现有建筑的基本几何构造，以及它的锯齿状金属结构和砖砌外墙。内部已被掏空，变成了一个文化中心，分为两个区域，有独立的入口和循环通道，而且具有强烈的视觉和空间联系，并有可能对它们进行改造以适应不同活动的需要。

新建的中间区把老区和新区分离开来，凸显了现有建筑的特点，并在室外起到了有保护性的过滤作用。中间区域建立的视觉联系可以赋予空间体验上的变化。

在入口处设置了宽敞的公共空间，功能上类似划定的市场，可以召开会议、交流信息和举办展览，视觉上好像广场外围慢慢地扩展到建筑内部。采用了高科技的屋顶可以最大程度地吸收自然光和通风。

可持续发展和节能改造

设计师一直以可持续的方式对这座老工业建筑和被遗弃的兵营进行翻新，兼顾能源效率和可再生能源来获取系统的整合。

现有的砖立面和金属型材制成的屋顶结构都被保留了下来，同时又制作了一面新的混凝土结构板，通过热激活效应用于建筑的暖通空调系统。现有的屋顶桁架和它的金属支柱仍未完工，而室内其余部分均已结束。

地热可再生能源用于给建筑物供暖和制冷，空气地面交换器可以对主要的换风进行预处理。与使用传统系统相比，采用这种空调系统能大大降低该建筑最终的能源成本。

Daoíz y Velarde Cultural Center

As part of the Daoíz y Verlarde complex of former barracks the objective is to preserve the architecture; a representative sample of Madrid's industrial and military heritage.

From the start, the idea was to respect the basic geometry of the existing building, as well as its saw-tooth metal structure and the brick-built facade. The interior space has been emptied to create a container for the cultural center, which is divided into two areas with separate entry points and circulation areas, but with a strong visual and spatial connection between them, and the possibility of their adaptation to different types of events.

A newly created intermediate space separates the former container from the new uses to bring the character of the existing

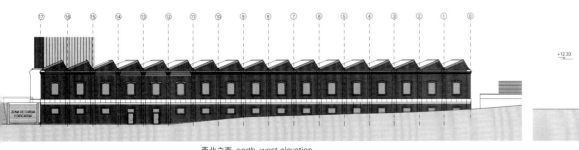

西北立面 north-west elevation

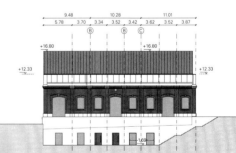

西南立面 south-west elevation

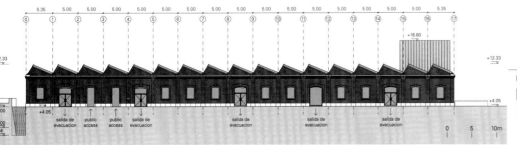

东南立面 south-east elevation

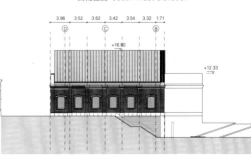

东北立面 north-east elevation

项目名称：Daoíz y Velarde Cultural Center
地点：Conjunto de los antiguos cuarteles de Daoíz y Velarde junto a la Avenida Ciudad de Barcelona, 162. Madrid
建筑师：Rafael de La-Hoz Castanys
合作建筑师：Rafael De La-Hoz Arquitectos
项目负责人：Silvia Villamor, Ángel Rolán
项目团队：Paola Merani, Concepción Cobo, Susanne Forner
施工技术员：Elena Elósegui, Javier Fernández / 平面设计：Daniel Roris, Luis Muñoz
模型：Fernando Mont, Víctor Hugo Coronel
工程指导：Rafael De La-Hoz Castanys
结构工程师：Ciete / 安装工程：Teisen / 施工单位：Férnández Molina Obras y Servicios S.A.
气候控制：Microclima / 地热能设计：Eneres / 照明设计：Antón Amán. Architectural Lighting Solutions
甲方：Madrid City Council, Área de Gobierno de las Artes, Dirección General de Infraestructuras Culturales
建筑面积：6,850m² / 施工时间：2007—2013
摄影师：©Alfonso Quiroga(courtesy of the architect)

building to the fore and set up a protective filter with the exterior. These intervening spaces establish visual connections that foster variations in the spatial experience.

A generous communal space has been created at the entry points; a place for meeting, information and exhibitions, which works as a covered agora, as if the square outside extended into the building. A hi-tech roof has been developed to take the best advantage of natural light and ventilation.

Sustainable and Energy Saving Refurbishing

The refurbishment of this former industrial building, and abandoned barracks, has been made in a sustainable way as regards energy efficiency and the integration of renewable energy capture systems.

The existing brick facade has been respected, as has the structure of the roof built from metal profiles, and a new structure of concrete slabs has been created, which will be used for the HVAC of the building through its thermo-activation. The existing roof truss and its metal pillars remained suspended in the air while the rest of the interior of the building was completed.

Geothermal renewable energy is used to heat and cool the building, and the air-ground exchanger works as a pre-treatment mechanism for the primary renewal air. The final energy cost of the building, employing this type of HVAC systems, is far lower than it would be otherwise using conventional systems.

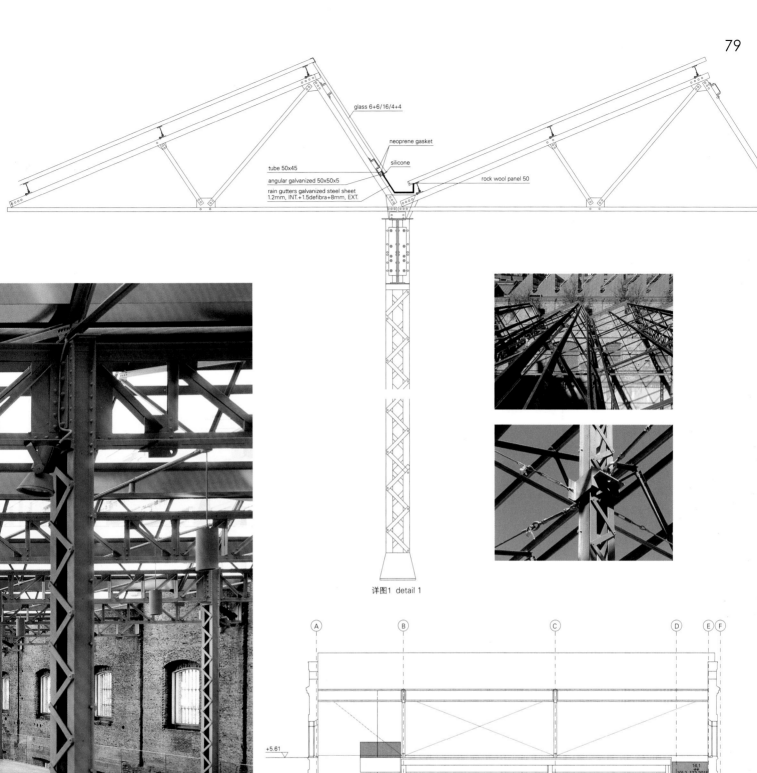

glass 6+6/16/4+4
neoprene gasket
silicone
tube 50x45
angular galvanized 50x50x5
rain gutters galvanized steel sheet 1.2mm, INT.+1.5defibra+8mm, EXT.
rock wool panel 50

详图1 detail 1

A-A' 剖面图 section A-A'

B-B' 剖面图 section B-B'

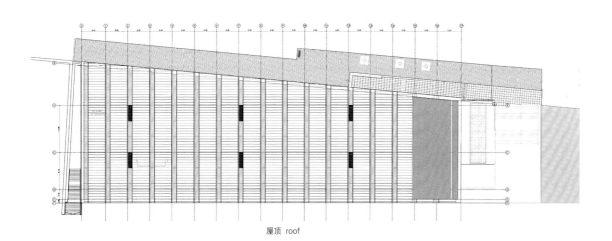

屋顶 roof

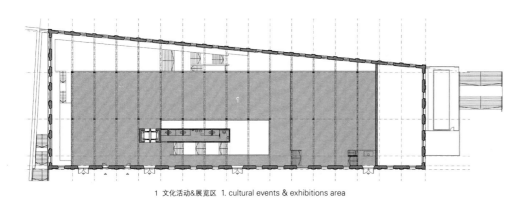

1 文化活动&展览区　1. cultural events & exhibitions area
三层　third floor

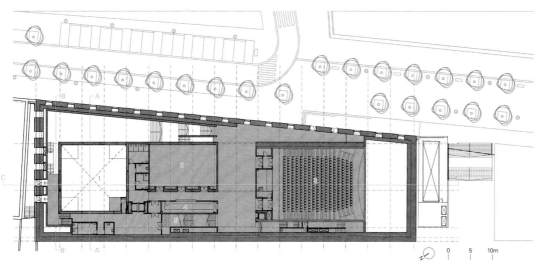

1 可配置的开放空间　2 工作坊&排练室　3 剧院&休息大厅&会议室　4 衣帽寄放处　5 仓库
1. configurable open space 2. workshop & rehearsal room
3. theater & lounge acts & conferences 4. checkroom 5. warehouse
一层　first floor

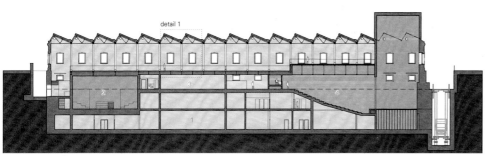

1 文化活动&展览区　2 可配置的开放空间　3 工作坊&排练室　4 剧院&休息大厅&会议室
1. cultural events & exhibitions area 2. configurable open space
3. workshop & rehearsal room 4. theater & lounge acts & conferences
C-C' 剖面图　section C-C'

Fjelstervang户外社区中心

Spektrum Arkitekter

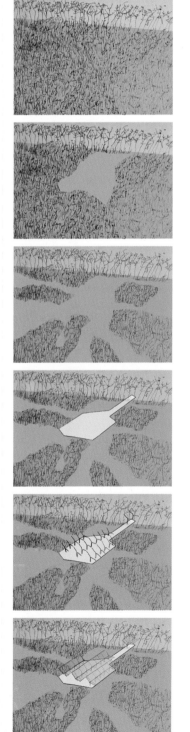

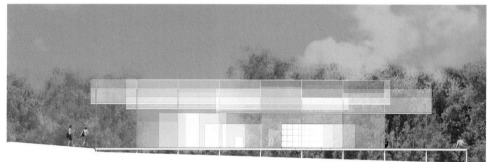

在志愿者的帮助下，500名生活在丹麦一个名为Fjelstervang的偏远小村庄的居民们经过团结努力，成功地修建了一所史无前例的半户外社区活动中心。作为"创建"活动的组成部分，在Fjelstervang建成的这个社区中心成了一个集会地点，在这里可以举办聚会、研讨会、教育活动、运动、舞会、圣诞晚会以及其他各种活动，不需要正式的房子，只要有个遮风挡雨的地方就可以；还可以烹饪和制作咖啡，以及围绕篝火团坐。

2013年春，Fjelstervang村得到了一些建筑师的协助，实现了建造半户外社区中心的梦想。利用问卷、研讨会、现场办公和开放式对话等方法收集了居民的意愿和观点，试图捕捉和理解居民愿望的本质进而把众多声音融入到一个简单而强烈的概念中。无论是否参与社区的各种社会活动和俱乐部，无论男女老少，这个概念足以让每个人都能实现自我认同，也都能感到宾至如归。

设计目标是建造一个通透、开放的娱乐中心，没有明显的室内和室外界限区分。朴素的选址、特别的建筑设计，让人们可以与周围的风景和自然进行交流。社区中心的建筑十分坚固，穿着跑鞋和球鞋的人们都可安心进入。建筑物的四周由轻便灵活的滑动墙构成，形成一种身处野外的感觉。整个建筑好似一个简易坚固的气候屏障，即使气候寒冷、刮大风，一年四季也都可在户外举办活动。

这个中心由并排的两个体量组成。透明的屋顶置于一排椽/柱子之上，高达6m。彩色的滑门白天可以过滤光线，晚上让整个建筑宛如发光的彩灯，不禁让人想起附近刚被关闭的染料厂。建筑的两个体量均建立在巨大的木板平台之上，约有500m²，营造了一种统一的表面效果。此外，该建筑还包括一间简易厨房和开放的炉子。所有这些元素共同促成了一个位置特殊、经历丰富、活动形式新颖的建筑结构。正如一位当地居民所说的："这一天标志着Fjelstervang村新纪元的开始。"

Fjelstervang Outdoor Community Hub

Thanks to unique unity, hard work, volunteer help and tremendous commitment, the inhabitants of Fjelstervang, a small rural village of 500 people, have succeeded in building a recreational semi-outdoor community hub in record time. As part of the campaigne "Build it up", Fjelstervang has now obtained a community hub that can be used for parties, workshops, education, exercise, folk dance, Christmas events and countless other activities that do not require a proper house, but can take place as long as there is shelter and possibility for cooking, making a cup of coffee and sitting together around an open fire.

Spring 2013, Fjelstervang Village won the assistance of architects to realize their dream of building a semi-outdoor community hub. From the inhabitants' wishes and ideas collected through questionnaires, workshops, on site office work and open dialogue, we have attempted to catch and understand the essence of the inhabitants' wishes and input and encapsulate the many voices in a simple and strong concept, a concept where, young and old, women and men can identify themselves and where all the inhabitants of the village can feel welcome regardless of whether one is connected to the various societies and clubs in the village or not.

1 厨房	6 休息室	1. kitchen	6. lounge
2 移动厨房	7 壁炉	2. mobile kitchen	7. fireplace
3 户外集会场所	8 大会议室	3. outdoor gathering	8. large chamber
4 午后露台	9 架子	4. afternoon terrace	9. shelves
5 上午露台	10 坐卧两用长椅	5. morning terrace	10. daybeds

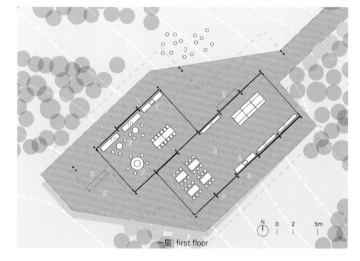

一层 first floor

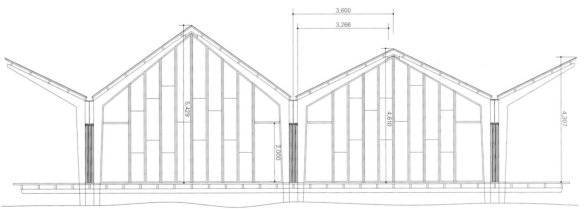

西南立面 south-west elevation

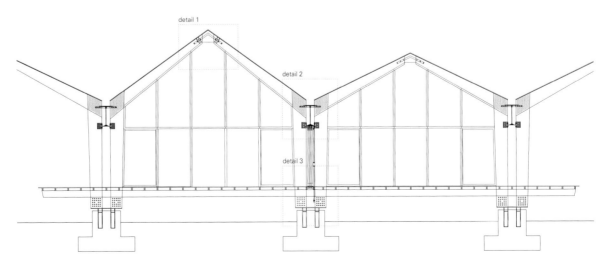

A-A' 剖面图 section A-A'

The ambition was to create a light and open recreational building where the border between indoors and outdoors is fluid. An understated and site-specific architecture is in dialogue with the surrounding landscape and nature. The community hub is a rugged building where you can, with a good conscience, enter both with running shoes and football boots. The building's sides consist of mostly sliding walls that are light and flexible and give an impression of being out in the open. Thus the building emerges as a simple and rugged climate shield, that extends use of the seasons and allows for events out in the open despite cold weather and wind.

The Hub stands as two volumes side by side. A transparent roof folds across a row of rafters/columns, about 6m high. Sliding doors in different colors, a reference to the recently closed textile dyeing factory, filter light during the day and make the building appear as a glowing and colorful lantern during the night. The two volumes on a large wooden deck, about 500m², have created a unified surface. Additionally the building consists of a simple kitchen and an open stove. Together these elements create a site-specific building that tells stories and creates a framework for new forms of life and activities. As one of the inhabitants of the village says at the inauguration: *"This day marks day one of the new history of Fjelstervang."*

项目名称：Fjelstervang Udeforsamlingshus / 地点：Fjelstervang, Vestjylland, Denmark
建筑师：Spektrum Arkitekter
项目团队：Sofie Willems, Joan Raun, Lærke Sophie Keil
合作方：Danish Broadcasting Corporation(DR), Realdania, The Danish Foundation for Culture and Sports Facilities(LOA), COWI, Danish Architecture Centre and Danish Arts
功能：recreational community hub
用地面积：500m² / 建筑面积：220m²
覆盖面积：roof_420m² / terrace_540m²
设计时间：2013 / 竣工时间：2013
摄影师：courtesy of the architect

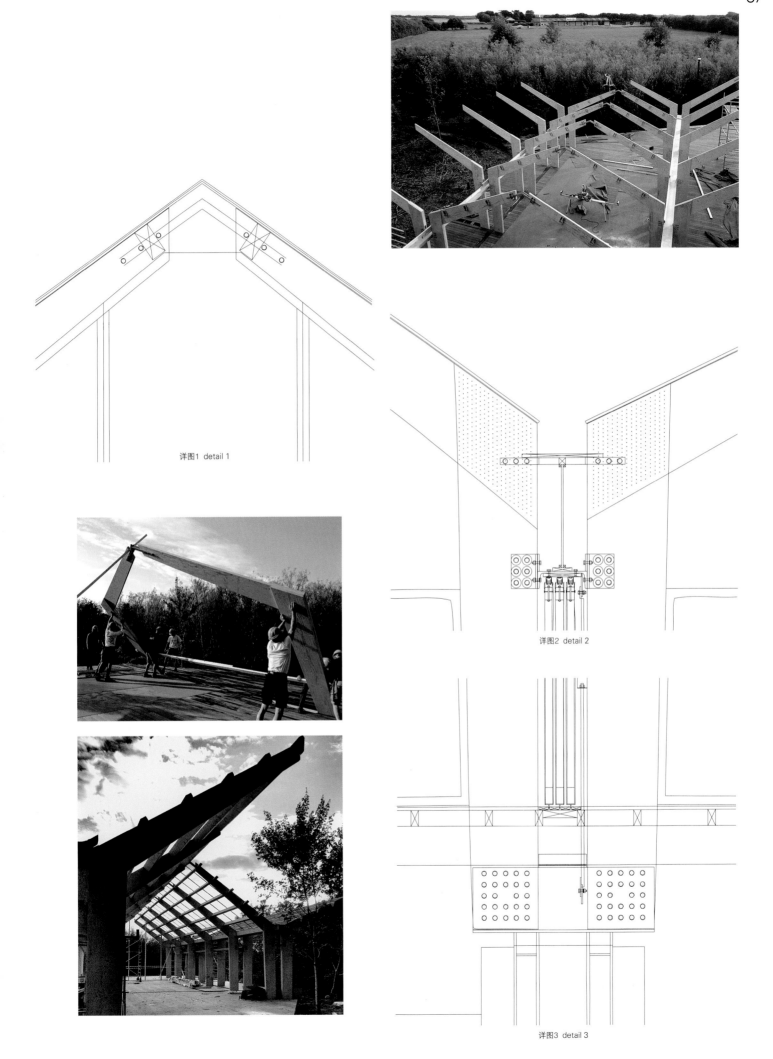

详图1 detail 1

详图2 detail 2

详图3 detail 3

友好中心

Kashef Mahboob Chowdhury/Urbana

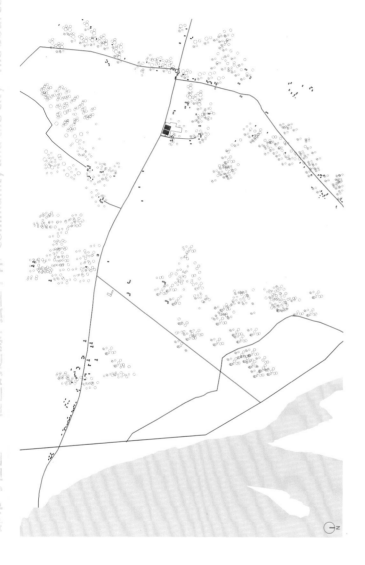

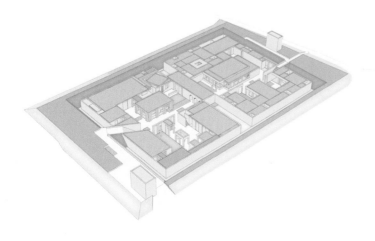

坐落于孟加拉国戈伊班达县附近的友好中心服务于一个非政府组织，该组织主要与这个国家最贫穷的人和生活在大河流域、缺少生活资源和机会的人们打交道。友好中心通过其建筑设施来进行自己的相关培训项目，同时也将这些设施租出去用于会议、培训等用途，以此来增加中心的收入。

中心所处的低洼土地位于戈伊班达县的乡间。该地区以农业为主，四面是环城的堤坝，因此，一旦决堤便有遭受洪灾的危险。

中心的项目广泛，资金却短缺，这意味着将中心的建筑提高到洪灾的水位（约为2.4m）之上并不是一种解决方法：几乎所有的可使用的资金加在一起也难以应付。因为处于地震多发区，当地土地承载力不足，也增加了建筑结构的复杂性。第三种以及最后一种设计依托周边的堤坝来防洪，因而直接以承重砌体的形式建在现有的土地上。雨水和地表流水被收集在内部的池子中，多余的水则通过压力泵输送到挖掘的池塘中去，用于渔业。设计依赖于自然的通风和冷气，庭院、水池和屋顶覆盖的泥土都促进了这种设计。四处设置的水箱与深井确保了污水不会混入其中。

"Ka"建筑体块包含接待亭、办公室、图书馆、培训/会议室和各种亭子，还有一个祷告区和一家小茶店。"Kha"建筑体块是由三个拱形门廊连接在一起的，多为私人功能，包括宿舍、食堂和员工及家庭区域。洗衣房和烘干室坐落在池塘的另一边，这里没有空调设施，全部照明都是通过LED和节能灯来实现的。至于建筑，正如刚开始设想的那样，中心的建设是对废墟的一种效仿，充满着对60公里外的Mahasthan（公元前3世纪）的遗迹的记忆。整个建筑从建设到完工主要依赖于一种材料——当地手工制成的一种砖——建造出亭阁、院落、水池和绿地；以及长廊和

隐蔽处。简洁的设计意图体现出了修道场所般的感觉。

中心服务于这个国家乃至全世界范围内的一部分极度贫困的人们，他们的生活依然很艰苦，即使是灯光、狭小的生活空间也成为一种奢侈，更别提运动和发现的乐趣了。

Friendship Center

Friendship Center near the district town of Gaibandha, Bangladesh, is for an NGO which works with some of the poorest in the country and who live mainly in riverine islands(chars) with very limited access and opportunities. Friendship Center uses the facility for its own training programs and will also rent out for meetings, training, conferences etc. as income generation.

The low lying land, which is located in rural Gaibandha where agriculture is predominant, is under threat of flooding if the embankment encircling the town and peripheries break.

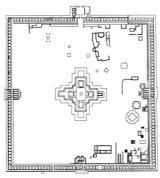

帕哈普尔寺, Naogaon
公元770—810年 (距离建筑基地85km)
Paharpur Monastery, Naogaon
AD 770~810 (85km from site)

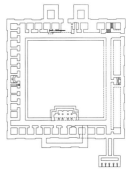

Sitakot寺, 迪纳杰布尔
公元7世纪—8世纪 (距离建筑基地110km)
Sitakot Monastery, Dinajpur
AD 7th ~ 8th C. (110km from site)

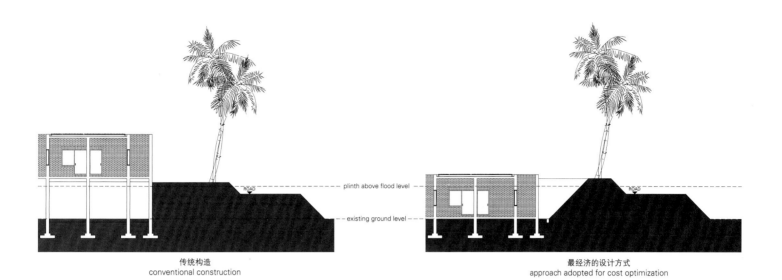

传统构造
conventional construction

最经济的设计方式
approach adopted for cost optimization

项目名称：Fridendship Center
地点：Gaibandha, Bangladesh
建筑师：Kashef Mahboob Chowdhury
合作建筑师：Anup Kumar Basak
项目建筑师：Sharif Jahir Hossain
项目团队：Irfat Alam, Raquib Al Hasan, Jewel Dewan
项目协调：Albab Yafez Fatmi, Sharijad Hasan
结构工程师：Matiur Rahman
电气工程师：Zafar Ahmed
暖通工程师：Phansak Thew
工程顾问：Jongsak Kuntonsurakan, KMA Bari
工地工程师：Nahidur Rahman
监理工程师：Amrul Hasan, Ahasanul Haque, Mohammad Ali, SM Hafizur Rahman, Jamir Ali Khan
用地面积：9,210m²
总建筑面积：2,897m²
设计时间：2008.5—2010.12
竣工时间：2011
摄影师：©Amrul Hasan (courtesy of the architect) - p.91
©Anup Basak (courtesy of the architect) - p.88~89, p.90
©Eric Chenal (courtesy of the architect) - p.92~93, p.94~95, p.96, p.97, p.98(except as noted)

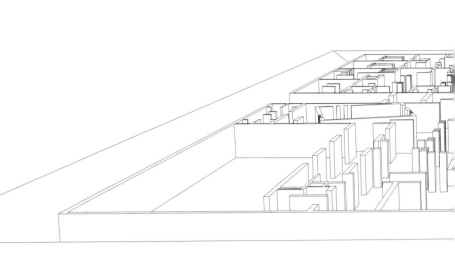

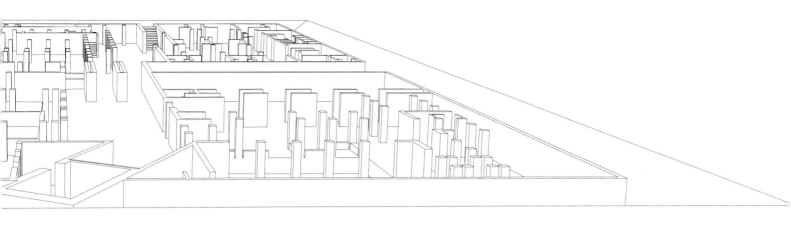

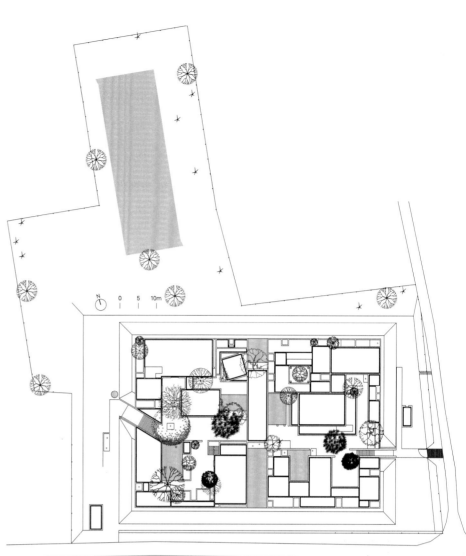

屋顶 roof

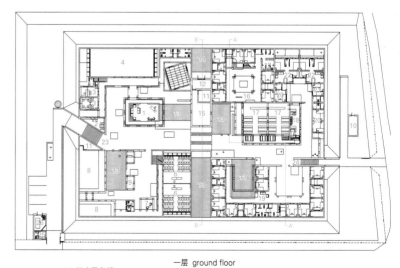

一层 ground floor

1 接待处	14 污水留存槽	1. reception	14. grey water retention tank
2 图书室	15 亭子	2. library	15. pavilion
3 办公室	16 女宿舍	3. office	16. female dormitory
4 会议厅(连着两间培训教室)	17 食堂	4. meeting hall (by joining 2 training rooms)	17. dining
	18 厨房		18. kitchen
5 祈祷空间	19 男宿舍	5. prayer space	19. male dormitory
6 培训教室	20 男员工宿舍	6. training room	20. male staff quarters
7 休息亭	21 女员工宿舍	7. breakout pavilion	21. female staff quarters
8 培训亭	22 家庭区	8. training pavilion	22. family quarters
9 就坐亭	23 入口	9. sitting pavilion	23. entry
10 净水设备	24 服务入口	10. water purification plant	24. service entry
11 储藏室	25 发电机	11. store	25. generator
12 水泵室	26 保安室	12. pump-out	26. security
13 地表水留存槽		13. surface water retention tank	

An extensive program with a very limited fund meant that raising the structures above flood level(a height of eight feet) was not an option: nearly the entire available fund would be lost below grade. Being in an earthquake zone and the low bearing capacity of the silty soil added further complications. The third and final design relies on a surrounding embankment for flood protection while building directly on existing soil, in load bearing masonry. Rainwater and surface run-off are collected in internal pools and the excess is pumped to an excavated pond, also to be used for fishery. The design relies on natural ventilation and cooling, being facilitated by courtyards and pools and the earth covering on roofs. An extensive network of septic tanks and soak wells ensure the sewage does not mix with flood water.

The "Ka" Block contains the reception pavilion, offices, library, training/conference rooms and pavilions, a prayer space and a small "cha-shop". The "Kha" Block, connected by three archways, is for

more private functions and houses the dormitories, the dining pavilion and staff and family quarters. The laundry and drying shed are located on the other side of the pond. There is no air-conditioning and the entire lighting is through LED and energy-efficient lamps.

As in construction, so in conception – the complex of the center rises and exists as echo of ruins, alive with the memory of the remains of Mahasthan(3rd Century BC), some sixty kilometers away. Constructed and finished primarily of one material – local hand-made bricks – the spaces are woven out of pavilions, courtyards, pools and greens; corridors and shadows. Simplicity is the intent, and monastic is the feel.

The center serves and brings together some of the poorest of poor in the country and – by extension – in the world, yet in the extreme limitation of means is a search for the luxury of light and small spaces; of the joy of movement and discovery in the bare and the essential.

南立面 south elevation

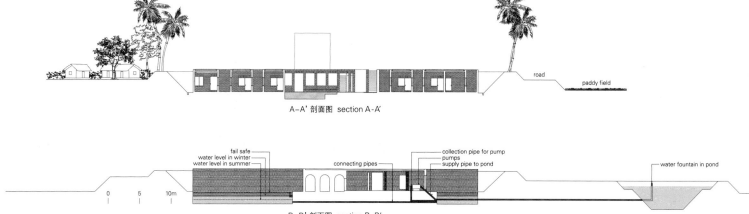

A-A' 剖面图 section A-A'

B-B' 剖面图 section B-B'

新Encants市场

b720 Arquitectos

Els Encants是一个具有百年历史的户外市集，一直以来采取的都是相对自由的交易模式。新建市场的位置与原有市场相距并不算远，都位于巴塞罗那的格拉利斯广场。

这个跳蚤市场用于拍卖和出售二手货、家具、服装和各种稀奇古怪的小玩意儿。它起源于14世纪。一开始是1929年在格拉利斯广场上占用了一个17000m²的公共场所用于临时的商业活动，后来渐渐在当时的基础上发展起来，成为深受人们喜爱的城市里独特的场所。

新建市场是一项重要的城市项目的第一步：改建格拉利斯广场，这座中央广场位于塞尔达三条主干路的交汇处。

几年来格拉利斯广场商圈的所有项目都提出过要挪走市场，因为市场的性质与新格拉利斯广场尊贵的市中心气息格格不入。但是Encants的商户强烈反对将这个市场搬离格拉利斯广场，他们的反对发挥了效用。

市政府最后决定将市场留在格拉利斯，但是将它挪到了广场对面的一个较小的场地。

虽然非常受欢迎，而且客流量很大，但是原来的市场场地设施不足，几乎没有任何投资，尽管是城里最能赢利的市场也不可避免地在走向衰落。

新设计试图保留原有室外交易市场开放的特性。但是，新项目受到场地的限制（只有8000m²），作为商业活动场所，面积至少应是目前的两倍多。所以这个项目无可避免地在某些程度上要堆叠起来。

为避免将其变成像普通购物中心那样的多层建筑，设计团队设计了一种连续的商业空间——利用微倾斜的平面形成能够连接小摊位还有商铺的循环空间，给到此购物的人一种在马路上闲逛的感觉。通过对平面的弯曲设计，实现了将新市场周边的不同标高道路协调引入。一旦进入市场，各个入口的层高就变得比较模糊。

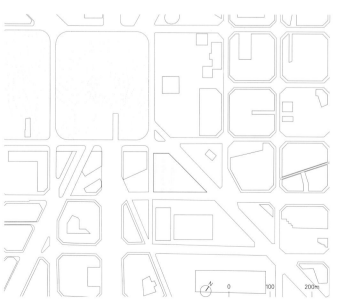

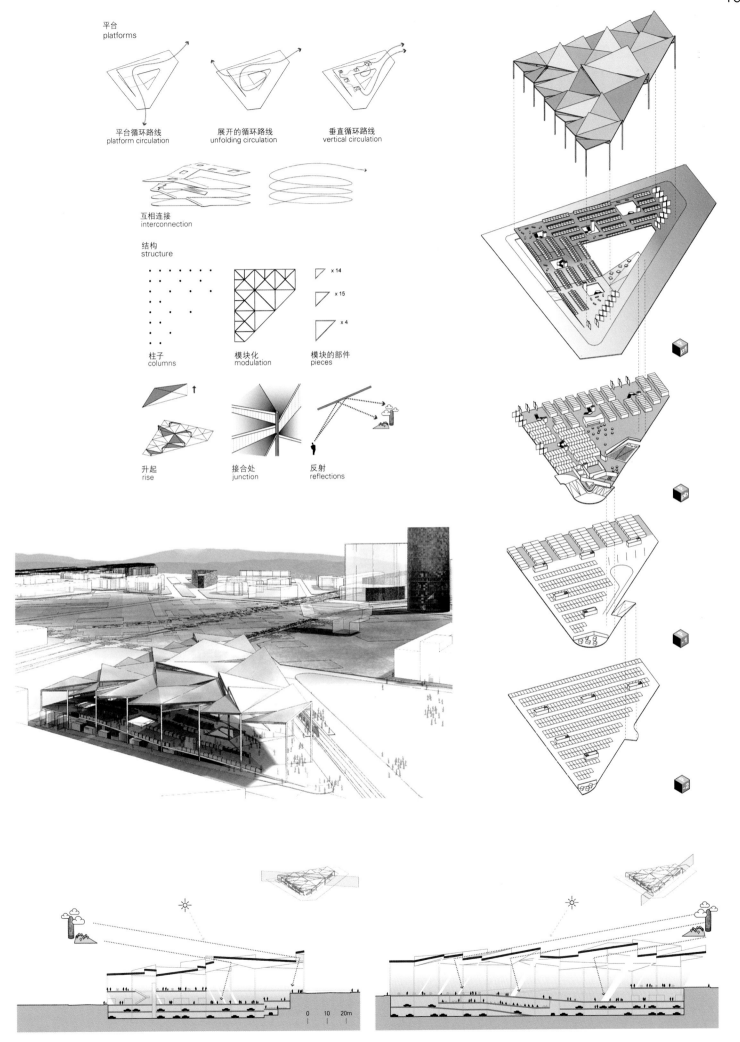

新公共空间的庄严之感和未来的格拉利斯广场的庄严之感是一样的，一个有着几百年历史的老公共建筑自豪地屹立在这里，让这种庄严之感更加实至名归。

屋顶成了视觉上的地标，它充满个性，人们在远处就可以辨认出来，非常引人注目。

尽管新市场算不上是建筑，它没有立面，但是其顶棚提供了巨大的空间，使它和附近庄严的国家大剧院建立起了和谐的对话。而且和未来的广场界限分明，不用牺牲市场的开放性优势。

新Encants市场的来客成倍增加。更重要的是，它在不排挤传统客人的同时，吸引了更多不同类型的造访者。这里甚至还成了旅游景点。

市场的品质得到了大幅度提升，拥有储存区、停车场、货物卸载区、行政管理办公室、会议室，最大化地减少了对市区的干扰，又不改变其原有的开放的市场氛围。

但是在所有令人引以为豪的特色当中，最重要的是，新Encants市场保持了它的灵魂。

New Encants Market

Els Encants is a centenarian marketplace set outdoors in an informal way. Its former location was close to the market's new grounds both at Glories Square.

A flea-market is dedicated to auctions and the sale of second hand objects, furniture, clothing and all sorts of curious and bric-à-brac. Its origin goes back to the 14th century. It began with its activity at Glories in 1929 occupying a public site of nearly 17,000m² in an improvised way, and developed under a temporary basis, becoming a much loved, unique space in the city.

The new market is the first step of an important city project: the reform of Glories Square, Cerda's central square at the junction of the three main avenues of this plan.

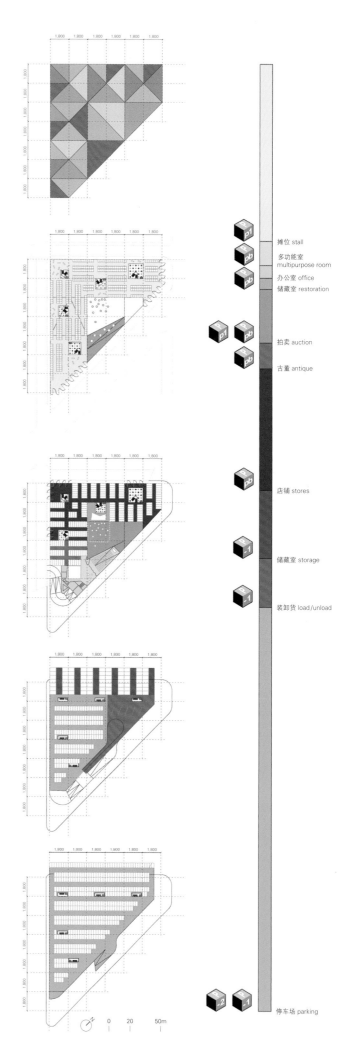

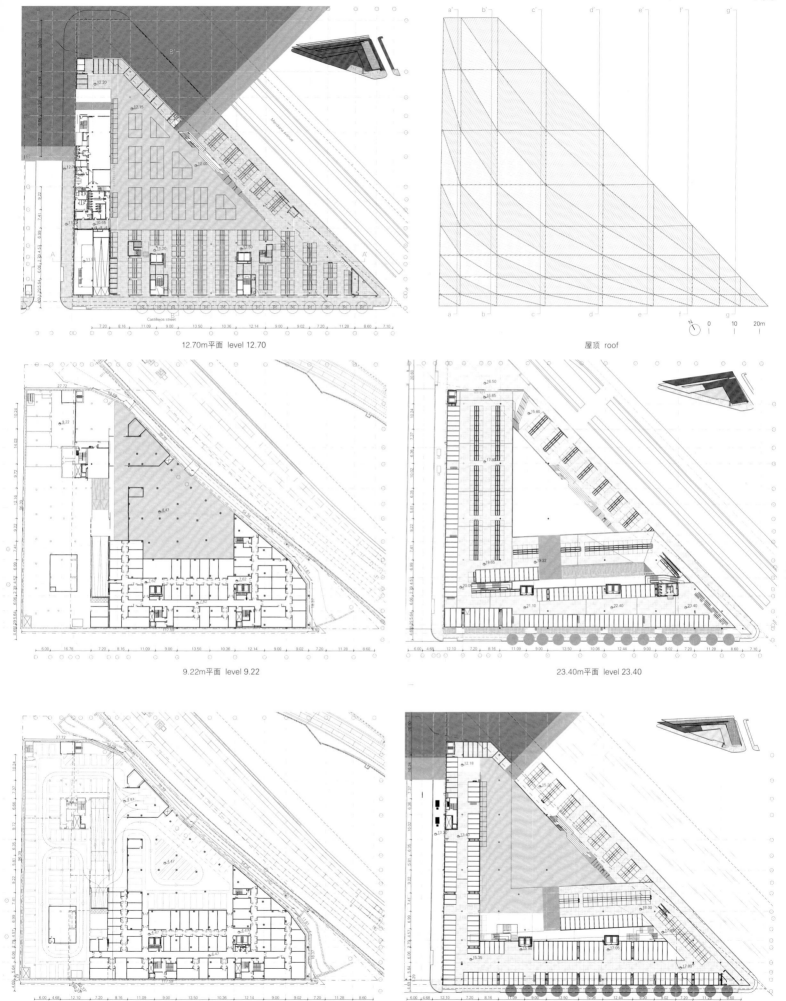

12.70m平面 level 12.70 屋顶 roof

9.22m平面 level 9.22 23.40m平面 level 23.40

4.47m平面 level 4.47 18.00m平面 level 18.00

Over the years all projects for the Glories Circus have assumed the removal of the market, as its nature has always been considered incompatible with the noble centrality meant for the new Glories. However, there has always been a strong and effective resistance of Encants merchants against any relocation that may take them away from Glories.

The city finally decided to keep the market in Glories but relocate it on a smaller land across the square.

Although popular and much visited, the previous grounds run down and had inadequate facilities with hardly any investment despite being the most profitable market for the city.

The main objective was to maintain the nature of the on-the-street outdoor market. The size of the new grounds(8,000m²) was a great constraint, as the commercial program was more than doubling that area. So the program had necessarily to be stacked in some way.

Its design intends to avoid multiple floors, rejecting a building model. Instead of that, a continuous area was designed taking advantage of the slopes of the surrounding streets. The street walkway peels off becoming a series of gentle ramps that climb on top of each other generating an endless loop. The visitor's experience is a stroll through a pedestrian road on top of which the stalls are arranged. The different levels of the streets around the perimeter are reconciled; therefore the various entrances are blurred.

The dignity of this new public space is equivalent to that of the future Glories Square and is deserved by a centuries-old public institution that will remain proudly there where they belong.

The roof creates a visual landmark that provides it with a strong identity, and makes it recognizable and inviting in the distance.

Despite not being a building and lacking facades, its canopy provides the volumetric entity that makes it possible to establish a proper dialogue with the neighboring solemn National Theater. It also creates a clear enough limit to the future Square without compromising its openness.

The new Encants Market has almost doubled the number of visitors. What is more relevant, it has broadened the type of people that visit it without expelling its traditional customers. It has even become a tourist destination.

It has improved dramatically the quality of its premises with storage areas, parking spaces, loading docks, proper management offices and meeting and assembly rooms, minimizing disturbances into the city without changing its typical open informal atmosphere.

But most important of all the new proud and vindicated, the new Encants Market has kept its soul.

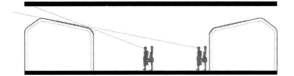

市场模块立面 market module elevation

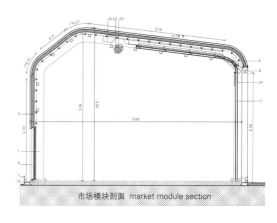

市场模块剖面 market module section

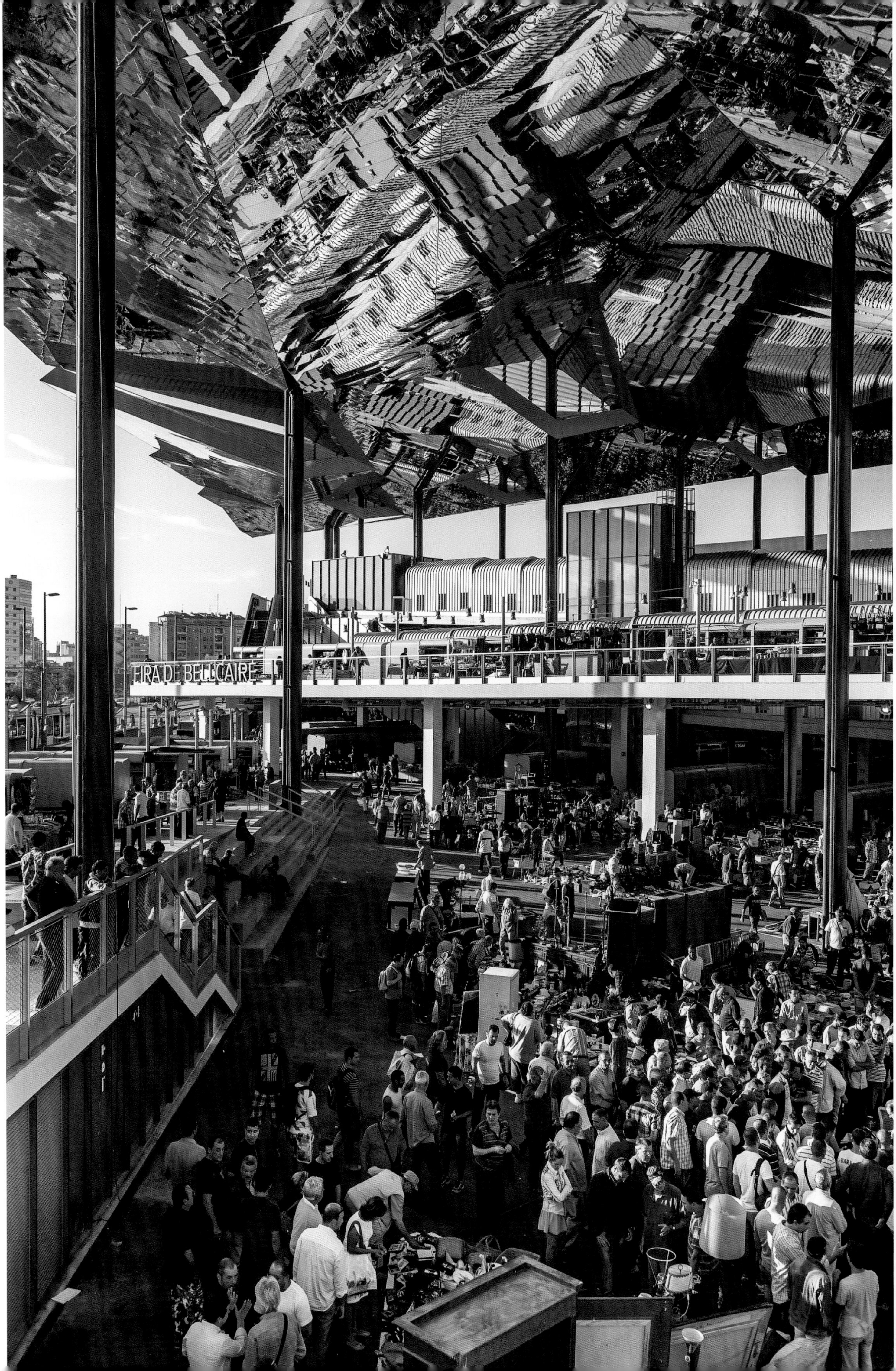
FIRA DE BELLCAIRE

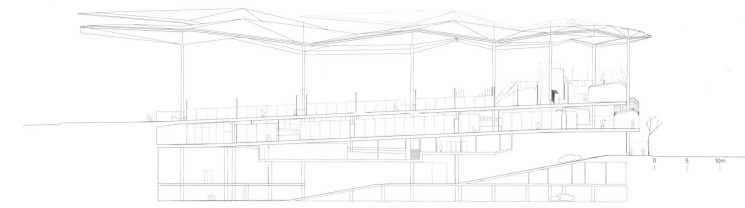

A-A' 剖面图 section A-A'

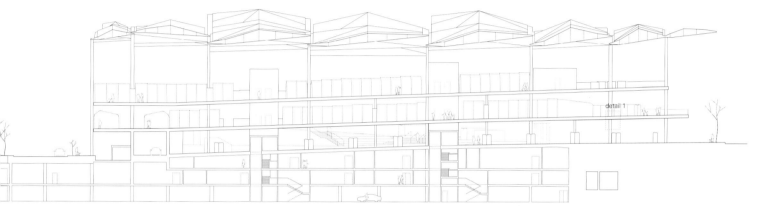

B-B' 剖面图 section B-B'

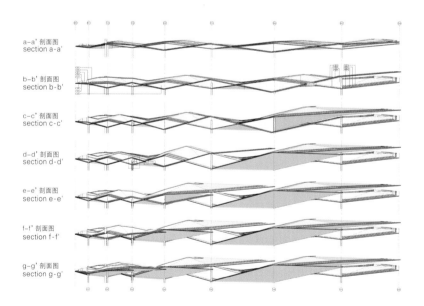

a-a' 剖面图 section a-a'
b-b' 剖面图 section b-b'
c-c' 剖面图 section c-c'
d-d' 剖面图 section d-d'
e-e' 剖面图 section e-e'
f-f' 剖面图 section f-f'
g-g' 剖面图 section g-g'

1. pillar formed by welded steel plates
2. truss formed by welded steel
3. supporting substructure perimeter shots formed by welded tubular steel
4. cover system consisting of formed trays with raised seal aluminum finish 0.9mm thick zinc
5. canalon plate steel with colaminado tpo membrane-FPA placed on media tableros water-repellent
6. rigid polyisocyanurate insulation
7. maintenance gateway welded grating trays attached to galvanized steel joints raised deck
8. trays cuffs folded composite panel with zinc finish, hidden fixing
9. transparent laminated glass louvers
10. tubular stainless steel structure for securing glass louvers
11. composite ceiling panels honey comb outer layer made of stainless steel AISI 304 mirror polished and colored, prepainted steel inner layer aluminum adjustable anchors
12. substructure fixing false ceiling
13. galvanized steel open profiles
14. downspout casting

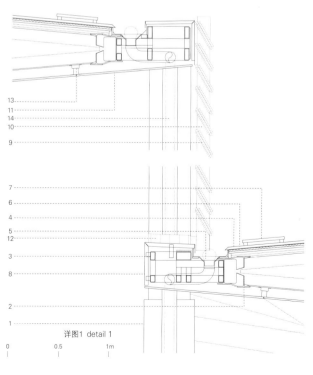

详图1 detail 1

项目名称：New Encants Market
地点：Plaza de las Glorias. Barcelona, Spain
建筑师：b720 Arquitectos
项目团队：Fermín Vázquez, Francesc de Fuentes, Cristina Algas, Sonia Cruz, Guillermo Weiskal, Pablo Garrido, Albert Freixes, Gemma Ojea, Javier Artieda, Angel Gaspar, Oriol Roig, Juan Pablo Porta, Leonardo Novelo, Jorge Mascaray, Francisco Marques, Helia Pires, Elies Porta, Tosca Salinas, Myriam Gonzalez, Nastascha Gergoff, Egbert Oosterhoff, Leopoldo Bianchini
结构工程师：BOMA
设备工程师：Grupo JG
甲方：Barcelona d'Infraestructures Municipals (BIMSA), Barcelona City Council
总建筑面积：35,440m²
造价：EUR 50,000,000
竣工时间：2008.10
设计时间：2008—2009
施工时间：2010—2013
摄影师：©Rafael Vargas - p.100~101, p.102, p.104, p.106~107, p.110~111 (except as noted)

建立社区的情景
The Scene that Builds

在城市中建立社区往往与场所的设计息息相关,场所即聚合人气、得到认可的空间,并在其中建立一种身份。在这里展示的建筑项目以提高人们对于常见的建筑形体和材料的可识别性为动力,以此让这些内容在建筑个体关系的发展中成为定位点。

17世纪末期,建筑师Jules Hardouin-Mansart设计了旺多姆广场,由于其钢铁般的立面,该广场已经成为一个极为重要的标志性区域,它设法展现了城市景观如何能够成为大众非常熟悉的区域,同时通过一个地方专有的共享感而把人们联系在一起。

由斯派克·琼兹导演的电影《她》,作为一种抽象且更加极致的认知处理,将讨论带回到时事中,作为对一些项目分析的导入,而在这里则可以通过三种可能的相关方法来展现:重复、雕塑和城市建设。

The establishment of communities within cities often goes hand in hand with the design of places that offer those aggregations of people recognizable spaces in which to develop an identity. The buildings shown here work by enhancing the recognizability of familiar shapes and materials to allow these to become anchor points in the development of relationships among individuals.
At the end of the 1600s Jules Hardouin-Mansart designed the Place Vendôme, which, thanks to the design of its steely facade, has become a place of extreme iconicity, managing to show how an urban landscape can become a territory of great familiarity, one that binds people through a shared sense of belonging to a place.
The movie Her, directed by Spike Jonze, as an abstract but more extreme treatment of recognition, returns the discussion to current events, serving as a lead-in to an analysis of several projects, here is seen through three possible approaches to the topic: repetition, sculpture, and city-building.

La Boiserie多功能活动中心_La Boiserie/DE-SO Architecture
布朗库堡区文化中心_Cultural Center in Castelo Branco/Mateo Arquitectura
Akiha Ward文化中心_Akiha Ward Cultural Center/Chiaki Arai Urban and Architecture Design
CIDAM农业经营研发中心_CIDAM /Landa Arquitectos
艺术广场_The Square of Arts/Brasil Arquitetura
比斯开新文化馆及图书馆_New Culture House and Library in Vizcaya/aq4 Arquitectura

建立社区的情景_The Scene that Builds a Community/Diego Terna

斯派克,Mansart,或重写社区

1986年,Mansart设计了巴黎的旺多姆广场,在此项工作中,他其实并没有设计一个广场,而是提出了关于城市的想法——被视为建筑的重写本,与住宅内部、商店和典型的城市功能相独立。Mansart设计的城市是一个建立在立面设计、比例和界定城市空间的表面装饰的基础上的地方。

Mansart所设计的建筑如同一个外部环境的目录册,以一分钟模式为特征,强调了集合都市的综合性景观:广场设计本身并没有大规模的设计策略——因为它没有考虑未来的场景,也没有形成系统的城市理念——它是非空间的,但是辨认感强烈的雕塑形式的对话。为了建造统一、可辨认、几乎是标志性的有机整体,如果说它有策略,那就在于描绘了每个俯瞰广场的单独的外立面图的行为。

Mansart用这种方法成功建立起了巴黎居民们对城市的同感,他设计的场景(一反常态的持久)能够被巴黎人和游客完全辨认出来。那些住在这里的人们能够感受到他们是同一设计的一部分,这种设计把每个单体建筑区分开来,而且把广场转变成一个巨大的内部空间:建筑师的作用就是把对空间限制的视觉化处理、细节的连贯性、广场周围精确描绘的实体和表面空间连接起来。

旺多姆广场是城市空间的一部分,并成为城市的特征,市民们经过

Spike, Mansart, or about a Community Palimpsest

In 1686, Jules Hardouin-Mansart designed the Place Vendôme in Paris. In doing so, he did not design a square, really, but an idea of the city – one envisioned as an architectural palimpsest, independent from the interiors of residences, shops, and typical urban functions. The city designed by Mansart is a place built on the design of the facades, on the proportions and the surface decorations that delimit the urban space.
The city of Mansart is a catalog of external environments, characterized by a minute pattern which emphasizes a comprehensive view of conurbation: The design of the square is not, in itself, a strategy on a large scale – it does not speculate on scenarios of the future, does not produce a systematic urban idea – it is a dialogue of sculptural forms, non-spatial, but strongly identifiable. Its strategy, if it has one, subsists in the very act of drawing every single facade that overlooks the square in order to build a unified organism, recognizable, almost iconic.
Mansart succeeds, using this approach, in developing an empathy with the inhabitants of Paris, designing a scene (paradoxically durable) that is fully recognizable for Parisians and visitors alike. Those who live here are able to feel they are a part of the same design, one that distinguishes each individual building and turns the square into a giant inner space: The role of the architect, then, is tied to his visual treatment of the limits of this void, to the consistency of detail, and to the precisely delineated solids and voids in the surfaces that surround the square.
It is in being part of a place thus characterized that citizens transiting through Place Vendôme can feel themselves as part of a larger

a Community

广场时能感受到他们自己是更大的集合体的一部分：简而言之，这个广场的丰富设计和高度可辨识性能够提升某种人际关系网络，而这种人际关系网络为找到自己的方位仅需要一个地方来定位。

在300多年后的2013年，我们在斯派克·琼斯导演的电影《她》中发现，当时城市建设的方法与Mansart在巴黎的建造方法相似，在强调个体之间关系的时期，不管这些关系是在人们之间或是在软件之间，好像已经有必要定义蕴含着视觉表象和设计的建筑背景。

斯派克·琼斯通过美国和亚洲建筑的结合体（尤其是上海浦东的金融区域）创建了洛杉矶的未来，选用艺术指导KK Barrett的话，给出了一个男人和虚拟程序之间不寻常关系的故事的完整感；它提供了电影角色和他们周围世界的更多信息，环绕周围的造型设计和他居住的泡泡也成为电影角色的一部分。（从la.curbed.com网站摘录）

电影的配景图包含了一些建筑的拼贴，这些建筑没有与旺达姆广场极端的统一性相呼应，但是确实设法建造了与众不同的景观：电影的配景图并没有使用现有城市或虚拟一个全新的城市，导演从其他城市的建筑物中借取并把它们融合在一起形成一个全新的地域。这些不是精确可辨认的物体，但是这些物体具有的必要特征使人们对其非常熟悉；仿佛斯派克·琼斯已经能够预测当代大都市的共同特征，他能够提升并以完全熟悉的方式把当代都市特征还原出来。结果，这座完整的城市成为可辨认的、几乎标志性的电影元素。

电影以人们之间的关系（电影中没有讲述孤独，而是描述了个体之间的革新性关系）展开了发展背景，这个背景成为强调和提升人与人之间关系的完美框架，融合了新居民社区的发展。

这里将要展示的项目描述了不同的社区，这些社区都与建筑支撑的社区发展相关。因为这些建筑具有很强的标志性，通过三种独特的方法使每个社区具有强烈的可识别性。

社区，三种背景
重复

2010年，智利的Elemental办公室对公益住房实施了研究，调查了在给定时间内社区居民自己建造一部分家园方面的定义。智利工作室仅设计了每间房屋的一小部分，为房屋自身的未来发展留下余地，剩下部分将由居住者自行建造。通过这种方式可能控制建筑的形式，但并没有完全控制居住者，留给他们自行设计的自由空间。在Verde别墅住宅项目的最后演化中，选取的系统展示了基本的且具有标志性并带有斜屋顶房子图形的广泛应用。这个影像与建筑的非连续性相反，这些建筑在中途停建，等待家庭的成长，保持了房屋的整体框架，人们可以有充分的自由来表达自己的想法。

aggregation: The square, in short, is able to promote, with its rich design and high recognizability, the kind of network of personal relationships that needs only a place to anchor in order to find itself.

More than 300 years later, in 2013, we find in the film *Her*, directed by Spike Jonze, an urban approach similar to that built by Mansart in Paris, at a moment when in addressing relationships between individuals, whether those relationships are between people or softwares, it seems to have become necessary to define an architectural background that has meaning in its visual imagery, in its design.

Spike Jonze builds a Los Angeles of the future through a composition of American and Asian buildings (particularly the financial district of Pudong in Shanghai), chosen, in the words of production designer KK Barrett, to give a sense of completeness to the story of an unusual relationship between a man and a virtual program: It gives more information about the characters and the world around them. The production design is wrapped around him and becomes part of his character – and the bubble he lives in. [excerpt from the site la.curbed.com]

The scenography of the film consists of a collage of buildings that do not communicate the extreme unity of the Place Vendôme, but do manage to construct a distinctive landscape: Instead of using an existing city, or imagining a totally new one, the director opts to steal architectural objects from other cities and bring them together to form a new territory. These are not precisely recognizable objects, but tend to have the necessary characteristics to make them extremely familiar; it is as if Spike Jonze has been able to extrapolate common characteristics of the contemporary metropolis, to exalt it and then return it in an absolutely familiar form. The result is an entire city that becomes a recognizable, almost iconic, element.

The relations between people in the movie (which tells not about loneliness, but about innovative relationships between individuals), unfold against a background that manages to become a perfect frame for emphasizing and promoting them, fostering the development of a community of new inhabitants.

The projects to be presented here tell the stories of different communities, all linked by how their growth has been supported by architecture that has such a strong identity that offers each community great recognizability, via three distinct approaches.

Community, 3 Backgrounds
The Repetition

In 2010, the Chilean office Elemental brought to a new fulfillment to its research on social housing, investigating aspects of the definition, at a given point in time, of a community of people who self-build part of their homes. The Chilean studio designs only a small part of each house, leaving room for the future growth of the house itself, to be built by its own inhabitants. In this way it is possible to have control of the forms of architecture, but not to fully harness the inhabitants, leaving them free space for self-design.

In the last evolution of the project, Villa Verde Housing, the chosen system shows intensive use of the figure house, elementary and iconic, with a pitched roof. This image is countered by a brutal in-

法国巴黎的旺多姆广场，Jules Hardouin-Mansart设计，1686年
Place Vendôme in Paris, France by Jules Hardouin-Mansart, 1686

Verde别墅，是智利的Elemental办公室设计的公益住房，2010年
Villa Verde, the social housing designed by Chilean office Elemental in 2010

因此在这种方式下，可能激活一种关于认知和归属的机理，且带有使独立性和丰富且混乱的发展相融合的可能性：这种融合有两种推动力，即规则和自由，它们的结合使空前的机会得到了难以置信的发展，使更多受限制的元素丰富起来，进而成为非常重要的元素。

Landa建筑事务所设计的CIDAM农业经营研发中心的主题是建筑元素的重复，使高层大楼具有辨识性。建筑结构使用了类似本土元素的形式并非是偶然的，尤其关于屋顶的轮廓、山墙的形式，不管是正常的还是相反的。

然而此项目让人回想起了现在每天的空间图像：它建造了一种集体回忆，把社区与空间本身连接，并因此能够将人们联系到一起。

重复的物体有一种人类的尺度，并且这种尺度在整个发展中反映出来：正如社区是个体的集合，建筑是小和中规模空间的集合。

雕塑

我们从瑞士巴塞尔抵达多尔纳赫。一座小村庄欢迎了我们，由遍及山区散落的房屋和共同的外观相连接组成：屋顶通常很明显；墙体外观类似液体水泥，软壁以固定的运转形式包围了内部。

这些房屋像是较大建筑的孩子，与周围富有特点的建筑的比例相比，这种感觉显得尤其强烈：由Rudolf Steiner于1923年设计的歌德堂成为人智学运动（是他于20世纪初期建立的）的中心。

歌德堂是一个巨大的建筑杰作，让游客叹为观止，吸引他们到达顶层，正如他们渴求到达山峰的顶点一样，如果不是这个巨大的混凝土体块，顶层也没那么重要。

像一尊巨大的雕塑一样，建筑与水泥的可塑强度相互交流，使一方想要围绕在另一方的身旁，由于这个庞然大物的存在，尽管是平凡的地方也仍然充满诗意。

由于这些特征和其尺寸与比例，歌德堂成为整个国家的焦点。不仅能从建筑方面而且还能从精神方面改变整个国家：好似城市规划只不过是建筑中央的球心，建筑野性的周围环绕混响，苍白仍具挑战性。

在焦点周围，体现人智学的社区建筑本身坚定地依附于歌德堂，使它充满信心地进入领域而无需恐惧分散，因为建筑将永远是一个安全的港湾。正是这种方式，Steiner对Mansart工作的成果产生极端的影响：不是建造一处空间上围绕社区故事的景观，这位奥地利设计师决定给予这种三维标志以生命，使它值得作为当地居民生活的定点。

Chiaki Arai城市和建筑设计事务所设计的Akiha Ward文化中心，Mateo建筑事务所设计的布朗库堡区文化中心和DE-SO建筑事务所设

terruption of the building, which stops in the middle, suspended in construction, waiting for the growth of the family, maintaining the overall profile of the house as a frame within which it has full freedom to express its own forms.

In this manner, therefore, it is possible to activate a mechanism of recognition and belonging that merges with the possibility of independence and fertile chaotic development: Two forces, rule and freedom, join to give rise to an incredible development of unprecedented opportunities that fertilize more restricted elements so they can become extremely vital ones.

CIDAM by Landa Arquitectos works on the theme of repetition of architectural elements that allow high building recognizability. It is no coincidence that the structure uses forms that resemble domestic elements, especially with regard to the profile of the roof, its gable form, whether normal or reversed.

Somehow the project hearkens back to images that are current in everyday spaces: It builds a kind of collective memory that links the community to the space itself and is therefore able to bind people.

The repeated object has a sort of human scale, and that scale is reflected in its overall development: Just as the community is the sum of individuals, the building is here the sum of small and medium-sized spaces.

Sculpture

We arrive in Dornach, Switzerland, from Basel. A small village welcomes us, comprised of houses scattered throughout the hilly area and linked by common appearance: a roof always evident; a runny appearance of the walls as of liquid cement; soft walls that wrap the interiors with crystallized movements.

These houses are the children of a larger building, immense when compared to the characteristic proportions of the surroundings: the Goetheanum, designed in 1923 by Rudolf Steiner, as the center of the Anthroposophical Movement, which he founded in the early 1900s.

The Goetheanum is a huge architectural masterpiece that takes the breath of visitors, attracting them to the top as they aspire to the summit of a hill that would be insignificant if on its top there was not this huge mass of concrete.

Like a giant sculpture, the building speaks to the plastic strength of the cement, makes one want to move around it, discovering a banal place that nonetheless becomes poetic due to the presence of this enormous object.

Thanks to these features and to its size and proportions, the Goetheanum manages to become a point of attraction for an entire country. It is able to transform it not only architecturally, but mentally: It is as if the urban planning of the place had been nothing more than a sphere centered on the midpoint of the building, with all around a reverberation, pale but no less challenging, of the building's brutality.

And around this focal point the Anthroposophical community builds itself, firmly anchored to the mass of the Goetheanum, which allows it to move into the territory with confidence, without fear of dispersal, because the building will ever be a safe harbor.

In this way, Steiner arrives at an extreme consequence of Mansart's work: Instead of building a scene, which spatially encloses

瑞士多尔纳赫的第二歌德堂，Rudolf Steiner设计，1928年
Second Goetheanum in Dornach, Switzerland by Rudolf Steiner, 1928

巴西圣保罗的SESC Pompéia项目，Lina Bo Bardi设计，1982年
SESC Pompéia in São Paulo, Brazil by Lina Bo Bardi, 1982

计的La Boiserie多功能活动中心都精确地分享了Steiner的意图：不仅让立面形成城市社区的背景，或者更好地，使其变成为三维背景，像斯派克·琼斯的电影中的一样，还要使其因为独特性成为标志性景观。

作为奇异性景观，这里展示了三个分析的项目，这些项目获得了区域的规模，成为非空间物体：它们的辨识性是直接的、有吸引力的；是社区和区域、陆地与居民之间的连接。在这种感知中，形式通常是单一的，而不能重复：作为伟大的艺术品，它更大的价值来源于它的独特性，也鉴定了参与其中人们的组别。

城市建设

Lina Bo Bardi的工作人员拍摄了一些图片，展示了位于圣保罗的SESC Pompéia项目，这里到处是享受日光浴的人们，穿着泳衣，在日光伞下，好像在沙滩上一样。正是1986年，由意大利的巴西籍建筑师设计了多功能中心，最近这位建筑师已经正式开始建造。

现代主义对首都巴西利亚的影响仍然强烈，圣保罗建筑为城市提供了惊喜：原材料、强烈的色彩、洞口的不规则性、非常规的比例为新建筑的景象创造了惊喜，这里注定要适应大量文化、社会和体育活动。

为了适应广泛的规划，高度分散且横跨不同功能，Lina Bo Bardi赋予了一组复杂的柱体以活力，来嵌入现有建筑，给它们增添了粗糙、强大的魅力，令游客为之惊讶。

人们能够感受到材料的粗野、体块的坚韧、光明和黑暗的扫掠，而且人们感受到非同寻常的舒适，感觉可以在这些形式中放松，找到与当地居民融洽的相处方式。

上面讨论的项目犹如没有大海的游泳者，仅是许多个体中的少数，他们感觉被这个城市新社区所吸引，这个新社区是大都市中的小村庄。

aq4建筑事务所设计的比斯开新文化馆及图书馆和Brasil建筑事务所设计的艺术广场和SESC Pompéia项目类似：它们通过多样的形式和材料激发自己。但是更重要的是，通过它们的空间系统、空间之间的关系和多样性规划，它们把自己建造成复杂的大城市区域中容纳的小城市：小规模城市，因此人们之间的关系更为安心。

像斯派克·琼斯一样，这些项目定义了由Mansart提出的一个城市未来的可能模型：一个建筑成就的未来，利用城市景观作为人类故事的背景，放大、培育并因此产生大范围的可能城市社区。

the story of a community, the Austrian designer decides to give life to a kind of three-dimensional mark worthy to serve as an anchor for the lives of the locals.

Akiha Ward Cultural Center by Chiaki Arai Urban and Architecture Design, Cultural Center in Castelo Branco by Mateo Arquitectura, and La Boiserie by DE-SO Architecture precisely share Steiner's intent: Not only do the facades form the background of the urban community, or better, become three-dimensional, as in the Spike Jonze's movie, they go on to become iconic, because unique.

The three analyzed projects are presented as singularities in the landscape, acquiring a territorial scale, becoming non-spatial objects: Their recognizability is immediate, attractive; it is the link between the land and the people, between the community and the region. In this sense, the form is always singular, it cannot suffer repetition: As a great work of art, its greater value descends from its uniqueness, which also identifies the group of people who attend it.

City-Building

There are some photos taken by the staff of Lina Bo Bardi which show the SESC Pompéia, in Sao Paulo, full of people sunbathing, with swimsuits and parasols, as if on a beach. It is 1986, and the multi-purpose center designed by the Italian, naturalized Brazilian, the architect has recently been inaugurated.

The influence of modernism on the new capital, Brasilia, is still going strong, and the São Paulo's building offers a shock to the city: The raw materials, the strong colors, the irregularity of the openings, the unexpected proportions, create a sense of surprise at the sight of the new building, destined to accommodate numerous cultural, social and sporting events.

To accommodate a broad program, highly dispersed across quite different functions, Lina Bo Bardi chooses to give life to a complex set of volumes that erode existing buildings, adding to them with a rough, powerful grace, leaving visitors stunned.

One can feel the brutality of materials, the relentlessness of the masses, the scanning of light and dark, yet one feels an extraordinary comfort, a sensation of being able to relax within these forms, which find an easy fit with the inhabitants.

The swimmers without a sea discussed above are only a few of the many individuals who have felt attracted to this place that defines a new community in the city, a small village in the great metropolis.

New Culture House and Library in Vizcaya by aq4 Arquitectura and the Square of Arts by Brasil Arquitetura move like the SESC Pompéia: They activate themselves through their great variety of forms and materials. But more importantly, through their system of dimensions, relations between spaces, and multiplicity of programs, they establish themselves as small cities within the complexity of the urban territory that holds them: They are cities on a smaller scale, and, therefore, offer reassurance for relationships with and between people.

As Spike Jonze does, these projects define a possible example of an urban future proposed by Mansart: a future designed by architecture, by an urban scene that acts as a background for human stories, amplifying them, fostering them, and thus generating a wide range of possible urban communities. *Diego Terna*

La Boiserie多功能活动中心
DE-SO Architecture

城市与城乡 建立社区的情景 Community and the City – The Scene that Builds a Community

　　"Boiserie"翻译成英语是"木结构"的意思,它是一个可以容纳1000人的多功能活动中心。该建筑坐落于法国普罗旺斯的葡萄种植带之中,附近整个区域都被雄伟的旺图山包围。这是一个木材与稻草砖结合的独特案例。

　　这个项目被分为两个部分,以此来减少体量感,减轻它对环境的影响。入口亭子是一个低矮的水平式建筑,它的灵感来自于该区域的阶梯式干砌石墙,在法语中被称作"restanque"。这个亭子朝向外面的风景开放,在它的身后是12m高的雪松板遮阳立面,从正面围裹着音乐厅的抹灰外墙。这层木板表皮的角度参考了旺图山的山坡,创造出了一种与会议厅的城市特性相互呼应的轮廓。

　　这层倾斜的滤光层是由银灰色的雪松条板制成的。雪松投射出充满生气的阴影,映照在大厅赭色的面板上。这种双层立面的设计创造出了深度感,以及光与色的交替变化,到了晚上,整座建筑的灯光在两层外表皮之间转换,建筑俨然变成了一只闪亮的灯笼。

　　音乐厅那不合常规的内部结构通过墙上的木板条和采用同样方式布置的吊顶与建筑的外表一唱一和。外部的木结构基本上就是将内部结构翻了过来,以此来表现出这种宏伟的体量。这种木质的氛围唤起人们对传统的法式社区会堂的记忆,与此同时,这也是各种不同的可能使用的结构形式中最理想的一种。这里安放了可折叠的坐椅,整个大厅可以采用自然光照,也可以完全将阳光遮挡在外。木材和稻草砖的使用营造出了多重的声效环境。

　　当然,整个大厅也透着一股木材的味道,自然清新。

Modene street

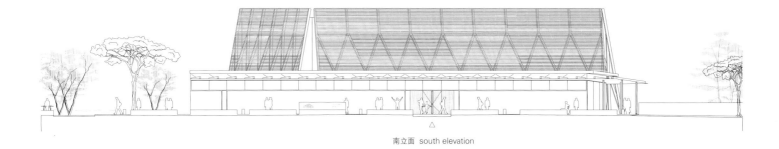

南立面 south elevation

La Boiserie

The "boiserie", which translates as the woodwork is a multifunctional event center with a 1000-person capacity. The building emerges from the striped viticultural landscape of Provence, France, and is dominated by the giant Mount Ventoux. It is a unique example of timber and straw-bale construction.

The program is split between two volumes to soften its impact on the surroundings and rupture its scale. The entry pavilion is a low horizontal building inspired by the region's dry-stone terraced retaining-walls called restanque in French. This pavilion opens out toward the landscape. In the background, a 12-meter inclined cedar shading facade protects the plastered exterior walls of the concert hall. The angle of the timber skin references the slopes of Mount Ventoux and creates a silhouette that communicates the civic nature of the meeting hall.

The sloping filter is made from slats of silver-grey cedar which project animated shadows onto the ochre-colored planes of the hall. This double facade creates depth and a vibration of color and light. At night, the building is lit between the two skins, transforming it into a glowing lantern.

The concert hall's atypical interior mirrors the exterior with the timber slats on the walls and ceiling using the same rhythm. The exterior wood is essentially turned inside-out to re-express the majestic volume. This timber ambiance evokes the traditional French community hall, and is ideal for a variety of possible configurations. The space is equipped with folding bleachers. It can be lit naturally or completely shaded from daylight. The use of wood and straw-bale insulation results in a versatile acoustic environment.

The space is infused by the smell of timber.

项目名称：La Boiserie
地点：Mazan, Provence, France
建筑师：DE-SO Architecture
项目经理：Matthieu Labardin
合作方：Nathalie Capelli
结构工程师：Gaujard Technologie
机械/电气/暖通工程师：MTC
声效：Altia
平面设计：FUGA
甲方：Mazan Townhall
功能：multi-purpose event space (615m², 640 seats), entry hall, exhibition space, bar, lounge, backstage changing rooms, activity room, exterior landscaping, parking
用地面积：1,740m²
建筑面积：1,480m²
总建筑面积：1,575m²
设计时间：2010
施工时间：2012
竣工时间：2013
摄影师：©Hervé Abbadie (courtesy of the architect)

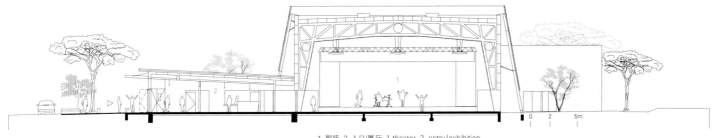

1 剧场 2 入口/展厅 1.theater 2. entry/exhibition
A-A' 剖面图 section A-A'

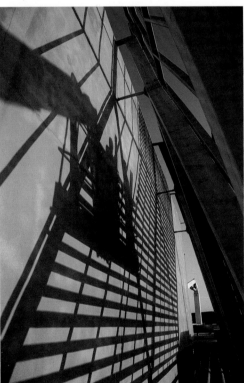
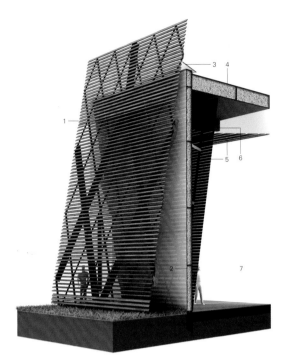
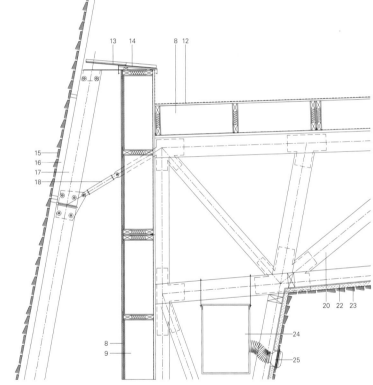
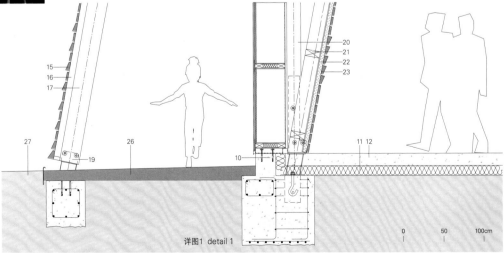

1. wood exterior skin
   - cedar wood structure
   - galvanized metal
   - wood louver with angle
2. vertical box made of wood
   - external stucco plaster
   - interior insulation with straw
3. glass protection
4. horizontal box made of wood
   - water proofing
   - interior insulation with straw
   - internal wood structure
5. acoustic wood ceiling
6. wood frame
7. theater
8. vertical box made of wood with straw
9. external stucco plaster
10. concrete
11. insulation
12. concrete polish finish
13. glass
14. metal structure
15. wood skin
16. galvanized metal structure
17. primary structure in cedar
18. bracing structure
19. metal structure
20. wood structure cedar
21. wood structure
22. interior acoustic absorption
23. acoustic panels in wood
24. M&E
25. blow air
26. stone
27. grass

详图1 detail 1

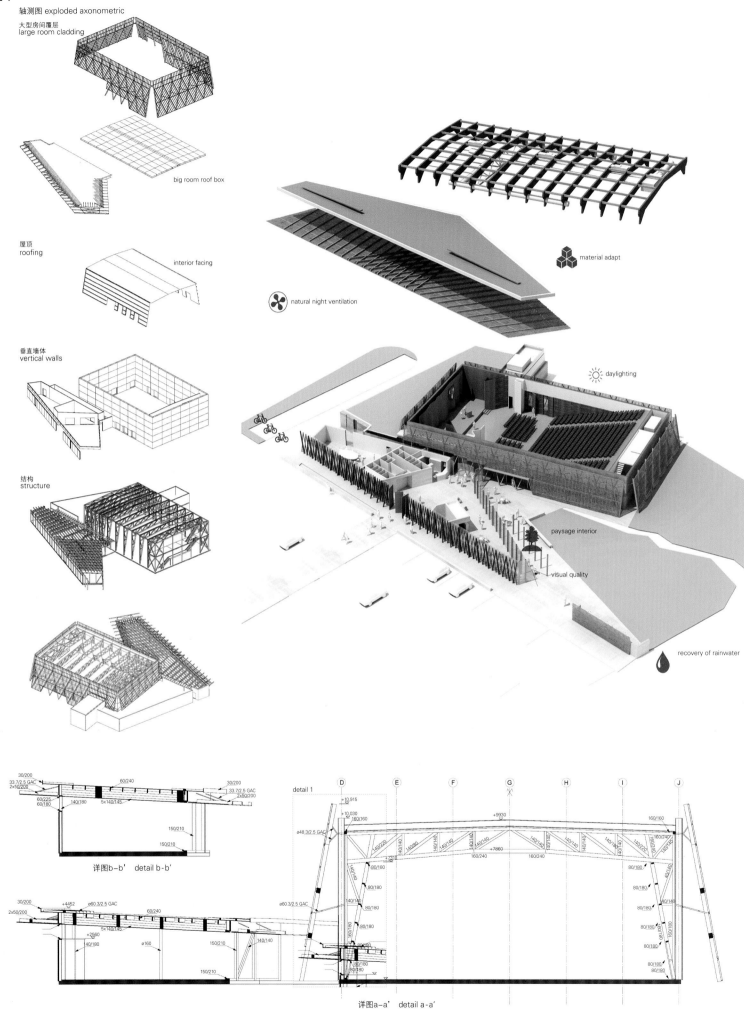

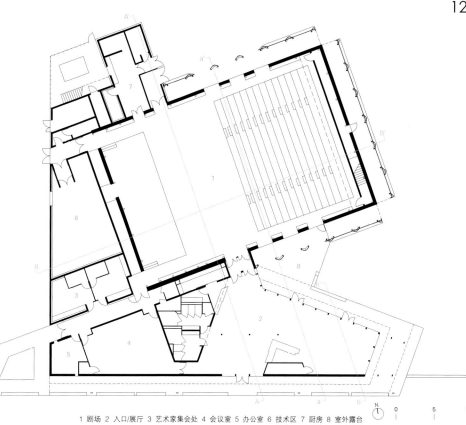

1 剧场 2 入口/展厅 3 艺术家集会处 4 会议室 5 办公室 6 技术区 7 厨房 8 室外露台
1. theater 2. entry/exhibition 3. artist lodge 4. meeting room
5. office 6. technical area 7. kitchen 8. exterior terrace
一层 first floor

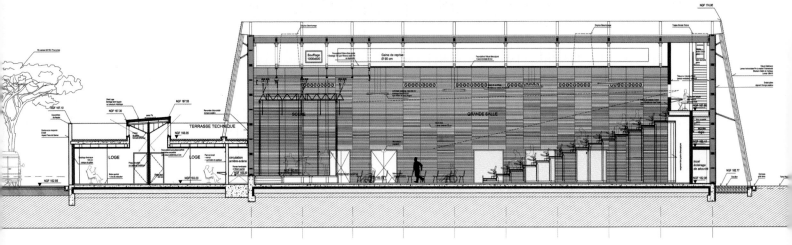

B-B' 剖面图 section B-B'

本项目表现出了历史悠久的布朗库堡文化中心复杂的公共空间、多样的交通方式和其他城市问题所带来的挑战。另外，建此文化中心的目的还在于要将这个老旧的城镇转变为城市的文化中心。

广场设计于项目一期（2007年），考虑到选址，需要解决最初的地形问题，同时在附近能建起其他各种建筑。广场坐落在山坡上，成交叉带状，依地势而建，可通往城堡。广场的缓坡自然而然地形成了在文化中心前方，广场中央的水池。文化中心建于项目二期，虽然是初始项目的组成部分，但却像桥一样，悬浮在广场上方的两个桩基础结构上，其底部就成了一个带顶的溜冰场，给这个巨大的公共空间、广场和附近的公园带来了连续性。溜冰场也是广场的一部分，利用了葡萄牙的溜冰传统和寒冷的大陆性气候。其木质立面与悬浮部分的锌板钢筋混凝土表面形成了对比，这里是一个活动区，悬浮在场址之上的屋顶和楼板连接了广场和公园。

从Praça Largo da Devesa大街折叠的铺砌路面形成的一个陡坡下来，可以到达布朗库堡文化中心的主入口。走近入口，从高处俯瞰的巨大的木条外立面让人眼花缭乱，它们的位置可以遮挡刺目的光线。沿斜坡而下，人们几乎被不知不觉地带到地下接待处，这里通往一个大画廊。这一层同时涵盖了行政区。

一条缓坡在建筑和广场地下延伸，延续了多种多样的楼层变化，将人们带入公共停车场。

在大楼里面，一层仅为过渡空间，与以上的楼层相连。而在外面，一层则是广场和文化中心的连接，容纳了贯穿建筑两端的溜冰场。溜冰场直接与周围环境相互作用，成为活动中心。这个户外空间让人愿意运动，充满了鲜艳的色彩，在夜晚提供照明，乐曲的旋律飘扬在空中。

采光天窗利用了一层的这个开口，将光线带入地下室，创建了明亮宜人的气氛。

通过一层木质立面以下的另一个入口，人们再次回到大楼内部。在较高的楼层，人们能看到模仿本建筑结构的一个礼堂和一个画廊，它们形成了双层高的空间。

在一端，展厅占据了二层和三层空间，并遵循建筑结构特点，利用斜坡来改变楼层高度。通过这种方式，参观者便可以纵观整个空间。

在另一端，礼堂利用座位的布置自然地融入了建筑的曲线。与台阶的浅色调相比，整体运用黑色更能抓住观众的注意力。

除此之外，一层还设有更衣室，可直达舞台。

二层的舞台对面是配电室和一个与礼堂主入口相连的供访客放松的酒吧。该层还设有多功能区，在展厅和礼堂之间呈半封闭状态。

在顶层可以眺望布朗库堡的壮丽美景。这座城市正是以此城堡来命名的。

最后，屋顶隐藏了所有的机器设备。在展厅上方开启了一扇巨大天窗，允许自然光进入。

布朗库堡文化中心

Mateo Arquitectura

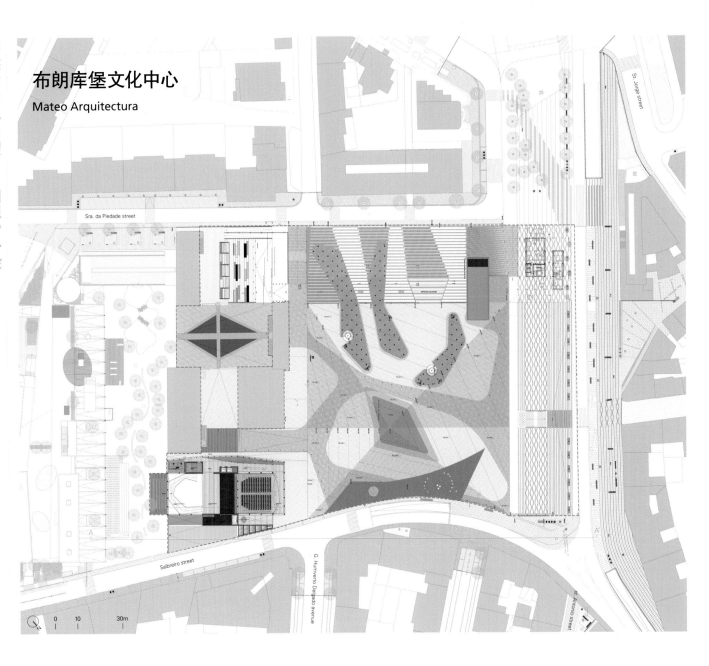

Cultural Center in Castelo Branco

The project presented the challenge of addressing the great complexity of the public space and the various traffic and urban problems of the historic center of Castelo Branco. The aim of the cultural center, furthermore, was to turn the old town into a cultural center for the city.

The plaza, designed in the first phase (2007), is moulded to the site to deal with the initial topographic problems and accommodate the various buildings designed to go there. Located on the slope of the hill that leads to the castle, it exploits the topography to form crosswise strips, giving rise in the central space of the project to a plaza whose gentle slopes give rise almost naturally to a pool of water in the center of the plaza, in front of the cultural center. The cultural center, built in the second phase, though part of the original project, floats on two piles over the plaza, like a bridge, freeing up at its base as a covered ice-skating rink, and giving continuity to this large public space, to the plaza and to the adjacent park. It forms another part of the plaza, drawing on the Portuguese tradition of skating and the cold continental climate. With its wooden facade, in contrast to the zinc-clad reinforced concrete of the suspended part, it is a bubble of activity, a roof and a floor float above the site, relating the urban sequence, the plaza and the park.

Descending one of the ramps generated by the folds in the paving of Praça Largo da Devesa, we come to the main entrance of Cultural Center in Castelo Branco. We move towards it, dazzled by the great facade of wooden slats, adjustable at one point to regulate the lighting, looking down on us from their position in the air. Almost without our realizing, this descent brings us to the reception, situated below grade and leading into a great gallery. This floor also accommodates the administrative area.

Continuing with the variations in the floor level, a gradual slope takes us to the car park, for the public, which expands beneath the building and the plaza.

Inside the building, the 1st floor is just a transition space, connecting with the floors above. On the outside, however, this floor is the manifestation of the connection between the plaza and the Cultural Center, housing an ice rink that extends from one side of the building to the other and interacts directly with its setting, becoming a hub of activity. It is an outdoor space that generates movement, color, light at night and music.

Skylights exploit this opening at 1st level to allow light into the basement floor, creating an ambience that is light and welcoming. We go back inside through another entrance on the 1st floor, beneath the wooden facade. On the higher levels, we find the au-

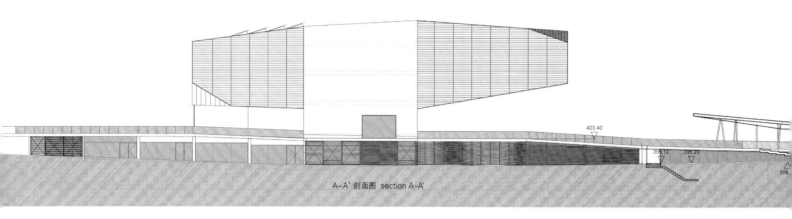

A-A' 剖面图 section A-A'

西南立面 south-west elevation

ditorium and a gallery that mimic the structure of the building, forming double-height spaces.

At one end, the exhibition hall occupies the 2nd and 3rd floors, with a ramp to change level that accompanies the structure of the building. In this way, the visitor has an overview of the space. At the other end, the auditorium also moulds naturally to the curve of the building with its seating arrangement. All in black, it contrasts with the lighter tones of the stage to focus the audience's attention.

In addition to these spaces, the 2nd floor also accommodates the dressing rooms, with direct access to the stage.

On the 3rd floor, opposite the stage, are the control room and a bar connected with the main entrance to the auditorium where visitors can relax. There is also a multipurpose space on this floor, enclosed between the exhibition hall and the auditorium.

The top floor offers stunning vistas of Castelo Branco, with the castle that gives the city its name.

Finally, the roof, concealing all the machinery, opens up in a great skylight over the exhibition hall, providing natural lighting.

Mateo Arquitectura

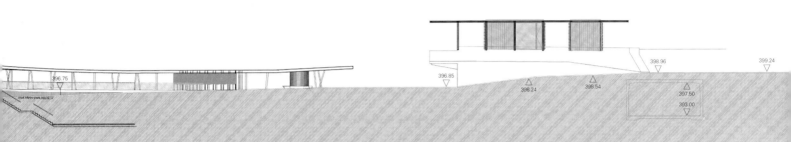

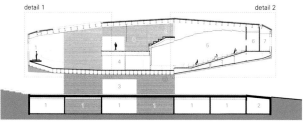

1 展厅 2 储藏室 3 溜冰场 4 更衣室 5 礼堂 6 控制室 7 门厅 8 门厅/展厅 9 放映室
1. exhibition 2. storage 3. skating rink 4. dressing room 5. auditorium
6. control 7. foyer 8. foyer/exhibition 9. projection
B-B' 剖面图 section B-B'

项目名称：Cultural Center in Castelo Branco
地点：Praça Largo da Devesa, Castelo Branco, Portugal
建筑师：Mateo Arquitectura
作者：Josep Lluís Mateo
合作方：Carlos Reis Figueiredo
甲方：Castelo Branco Council
用地面积：public space _ 60,000m², cultural center _ 4,300m²
造价：public space _ EUR 3,000,000
cultural center _ EUR 9,000,000
设计时间：2000
施工时间：public space _ 2006, cultural center _ 2012
竣工时间：public space _ 2007, cultural center _ 2013
摄影师：©Adrià Goula (courtesy of the architect)

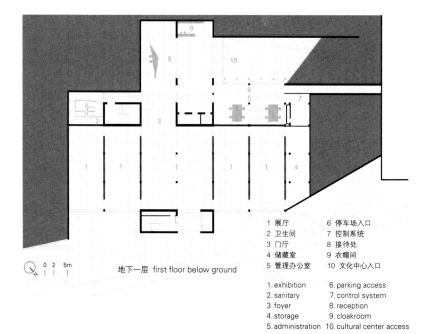

地下一层 first floor below ground

1 展厅　　　　6 停车场入口
2 卫生间　　　7 控制系统
3 门厅　　　　8 接待处
4 储藏室　　　9 衣帽间
5 管理办公室　10 文化中心入口

1. exhibition　　6. parking access
2. sanitary　　　7. control system
3. foyer　　　　 8. reception
4. storage　　　 9. cloakroom
5. administration 10. cultural center access

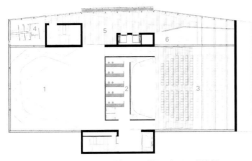

1 展厅 2 更衣室 3 礼堂 4 卫生间 5 门厅 6 设备间
1. exhibition 2. dressing room 3. auditorium 4. sanitary 5. foyer 6. technical room
二层 second floor

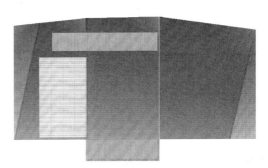

屋顶 roof

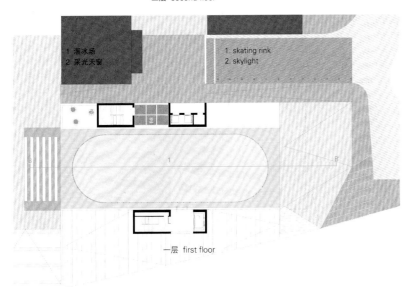

1 溜冰场
2 采光天窗

1. skating rink
2. skylight

一层 first floor

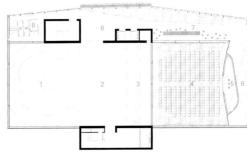

1 展厅 2 门厅/展厅 3 放映室 4 礼堂 5 控制室 6 门厅 7 门厅/酒吧 8 卫生间
1. exhibition 2. foyer/exhibition 3. projection
4. auditorium 5. control 6. foyer 7. foyer/bar 8. sanitary
三层 third floor

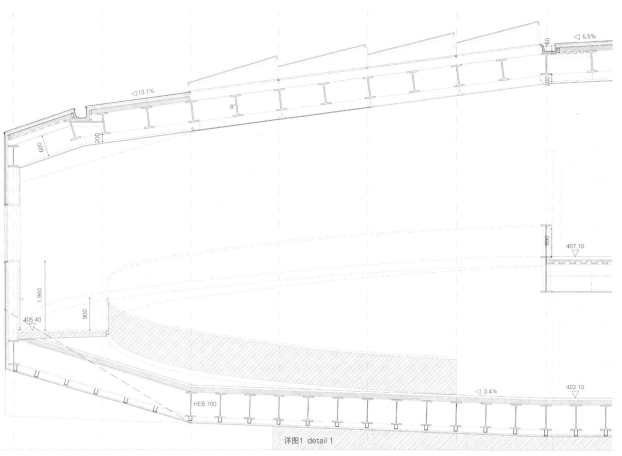

详图1 detail 1

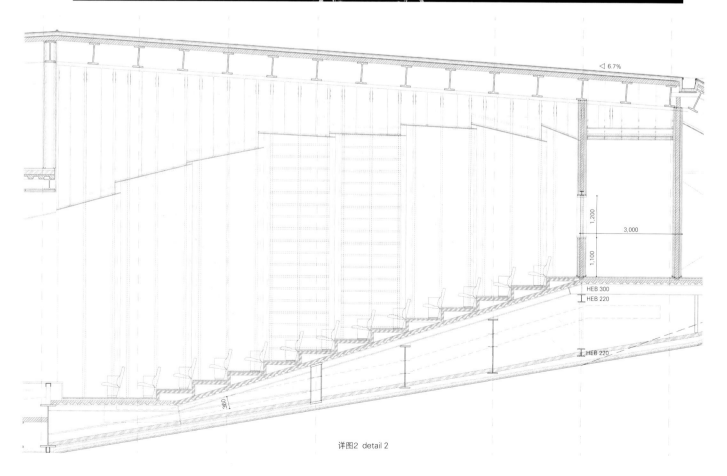

详图2 detail 2

Akiha Ward文化中心

Chiaki Arai Urban and Architecture Design

都市与社区——建立社区的情景 Community and the City - The Scene that Builds a Community

140

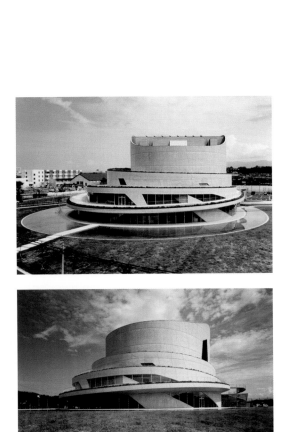

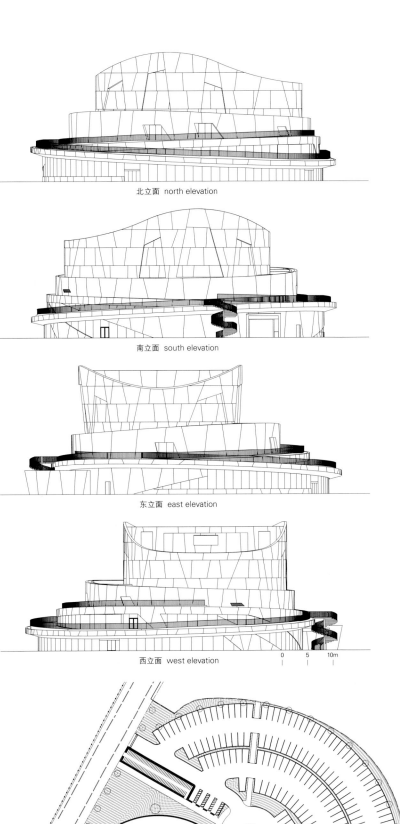

北立面 north elevation

南立面 south elevation

东立面 east elevation

西立面 west elevation

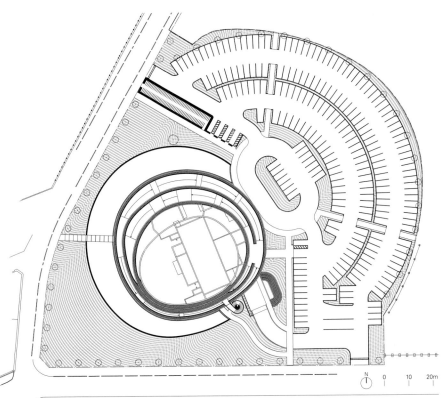

Akiha Ward文化中心位于新泻市著名的铁路工业区,面积为3000m²,拥有496个座位,被看作当地文化艺术的象征,民众对这座文化中心一直满怀期待。

坐落于17 000m²的原棒球场,建筑、风景和停车场沿着弧形的曲线唤起了人们对过去的回忆。周边是微微起伏的山丘地带,以及间或散落在平地上的居民区。保留山脉的原有特征,整体的建筑呈现了阶梯式的风景,人们在逐渐升高的过程中可以欣赏到周边的全景风貌。这个项目策划是在当地的工作间里研发的,应当地人的需求增加了一些房间和功能。在竞标阶段,设计师做了无数的改变,将外形从一个精致的圆形变成了由46个不同的弧组成的扭曲的圆形。规划分析图被分为不同的层次,依次为外部走廊、入口大厅、功能区、后台走廊和主大厅。为了使平面布局更加简洁高效,千秋新井将一些功能空间,如排练室、化妆间等用作入口走廊和后台通道,增加了可使用的空间。为了缝合整体外观和工作室平面规划之间的空隙,混凝土结构墙体被不断地弯曲扭转,从而达到了平衡地支撑建筑屋顶的效果。

主厅仿佛是掩在山丘下的洞穴。结构构件本身起到了反射声音的作用,因为这里的庞大厚实的体量发挥了超凡的声效效果。为了使声效效果最大化,这些混凝土结构体被看作一张巨大的网,在混凝土的网洞上安置穿孔铝板,用于吸音。大厅混凝土的内部结构为了声音的扩散完全都采用细凿琢面。加上灯光的效果,实心混凝土的外观有时显得庞大,有时显得轻盈飘渺,并且暗含着人手的温暖。

Akiha Ward Cultural Center

Located in a district famous for railway industry in Niigata city, Akiha Ward Cultural Center is a 3,000m² public theater having 496 seats. This building is designed to be a cultural incubator for the locals who have long-waited for it.

Sitting on a 17,000m² former baseball field, the structure, landscape and parking are organized along arcs to evoke the site memory. The surrounding area is a residential district on a vast flatland with a few small hills. Following in some characteristics of the hill, the global formation of the building is terraced landscapes where people can go up and take in the panoramic view.

The planning has been developed through workshops with the locals. Several rooms and functions were added in response to their requests. From the competition phase, the countless transformations changed the building's outline from a precise circle to a distorted circular form composed of 46 different arcs. The planning diagram is stratified in the order of exterior corridor, entrance lobby, functional rooms, backstage corridor, the main hall. Due to the simple diagram, several functional rooms such as practice rooms and dressing rooms can be used from both entrance lobby and backstage corridor, which improve operational availability. To accommodate the gap between the global formation and the planning based on the workshop, the structural concrete walls are bent and twisted to reach the balanced support points of the roof slabs.

The main hall is as if a concrete cave under the hill. The structural element itself works as acoustic reflectors which give extraordinary acoustics by the forth of its high specific gravity. To optimize the acoustic effect, the concrete structure is perforated like a net, and the porous aluminum sheets are installed in the holes as acoustic absorbent. Interior finish of the concrete structure of the hall is fully-dabbed for sound diffusion. With the lighting effects, the solid concrete looks sometimes massive, sometimes weightless, and it implies warmth of human hands.

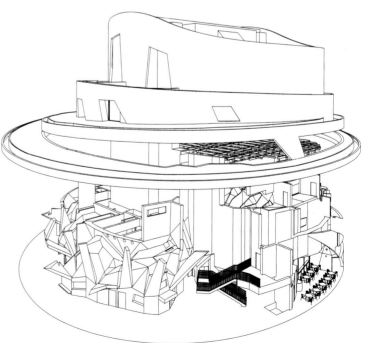

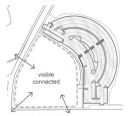
开放的场地规划
open-minded site planning

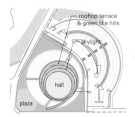
以绿色屋顶空间环绕
surrounded with rooftop green

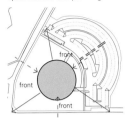
没有层次的形式
non-hierarchical form

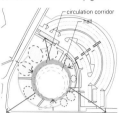
创造循环路线走廊
making circulation corridor

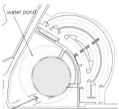
以水环绕
surrounded with water

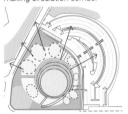
文化活动的扩展
expansion of cultural activities

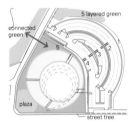
以地面绿色空间环绕
surrounded with ground green

场地规划
site planning

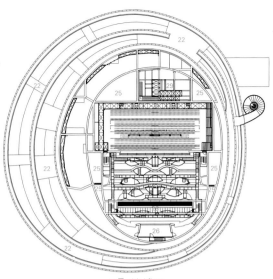
三层 third floor

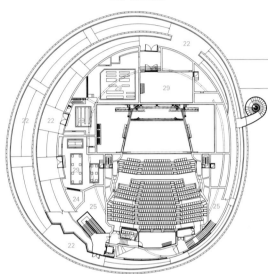
二层 second floor

1. water pond
2. cafe
3. kitchen
4. cloakroom
5. foyer
6. air chamber
7. stage
8. practice room
9. controlling room
10. recording studio
11. dressing room
12. entrance lobby
13. front desk
14. office
15. suckling room
16. meeting room
17. piano warehouse
18. warehouse
19. pantry
20. green room
21. service entrance
22. rooftop terrace
23. back stage corridor
24. family room
25. void
26. spotlight room
27. catwalk
28. gallery
29. machinery room

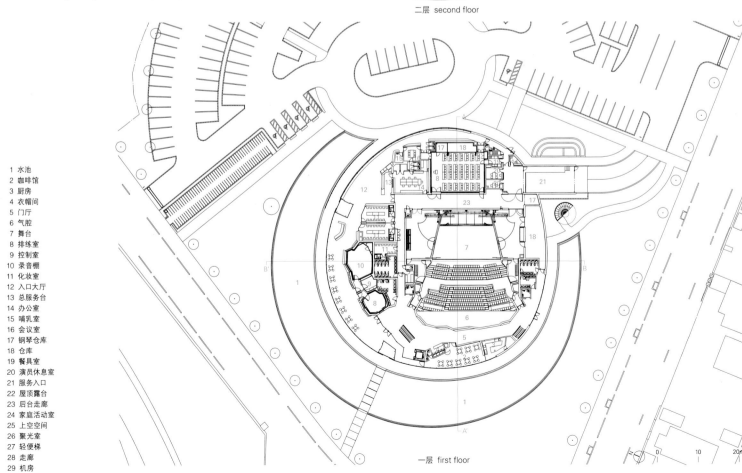

1 水池
2 咖啡馆
3 厨房
4 衣帽间
5 门厅
6 气腔
7 舞台
8 排练室
9 控制室
10 录音棚
11 化妆室
12 入口大厅
13 总服务台
14 办公室
15 哺乳室
16 会议室
17 钢琴仓库
18 仓库
19 餐具室
20 演员休息室
21 服务入口
22 屋顶露台
23 后台走廊
24 家庭活动室
25 上空空间
26 聚光室
27 轻便梯
28 走廊
29 机房

一层 first floor

结构参照物——卷心菜
structural reference_cabbage

在人们能接触到的地方，墙面是垂直的
walls stand perpendicularly where people touch

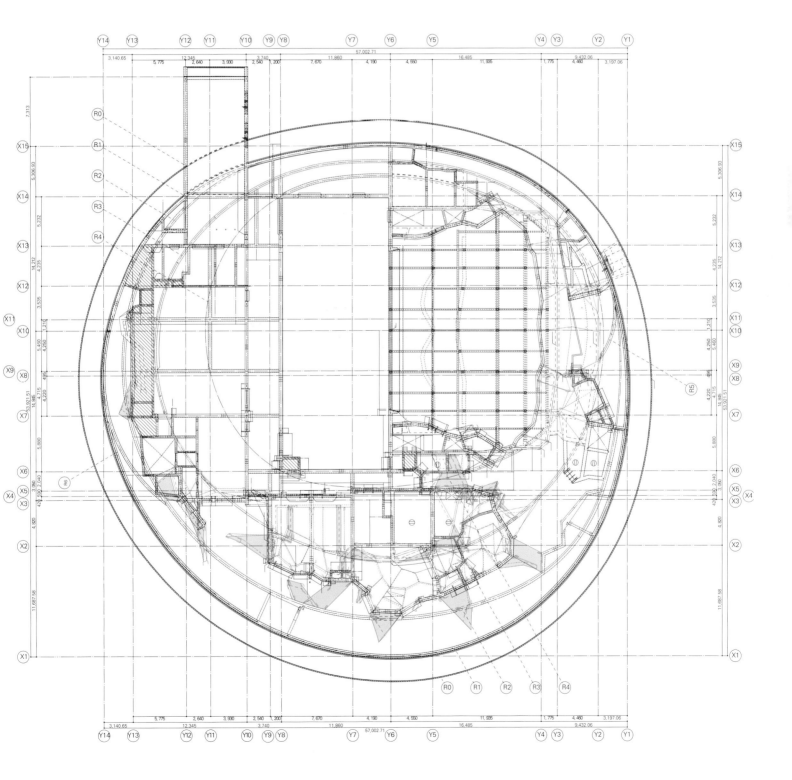

项目名称：Akiha Ward Cultural Center / 地点：4-23 Shineicho, Akiha Ward, Niigata city, Niigata, Japan
建筑师：Chiaki Arai Urban and Architecture Design
设计团队：Chiaki Arai, Ryoichi Yoshizaki, Tomonori Niimi, Akira Sogo
结构工程师：TIS & Partners Co., Ltd. / 工程与空调设备：SOGO Consultans
声学顾问：Nagata Acoustics / 剧院顾问：Theatre Workshop
室内与景观设计：Chiaki Arai Urban and Architecture Design
施工：Taisei and Shinko and Tanaka Joint Venture / 电力设备：Sakaden and Yae and Saito Joint Venture
空调卫浴设施：Niigata Kogyo Co.,Ltd. / 天然气设施：Echigotennen Gas Co.,Ltd. Niigata Branch
用途：theater (496 seats), practice room, studio
用地面积：17,165.34m² / 施工面积：2,848.86m² / 总建筑面积：2,997.36m² / 建筑规模：2 stories
设计时间：2009.11—2011.3 施工时间：2011.7—2013.5
摄影师：©Taisuke Ogawa (courtesy of the architect)

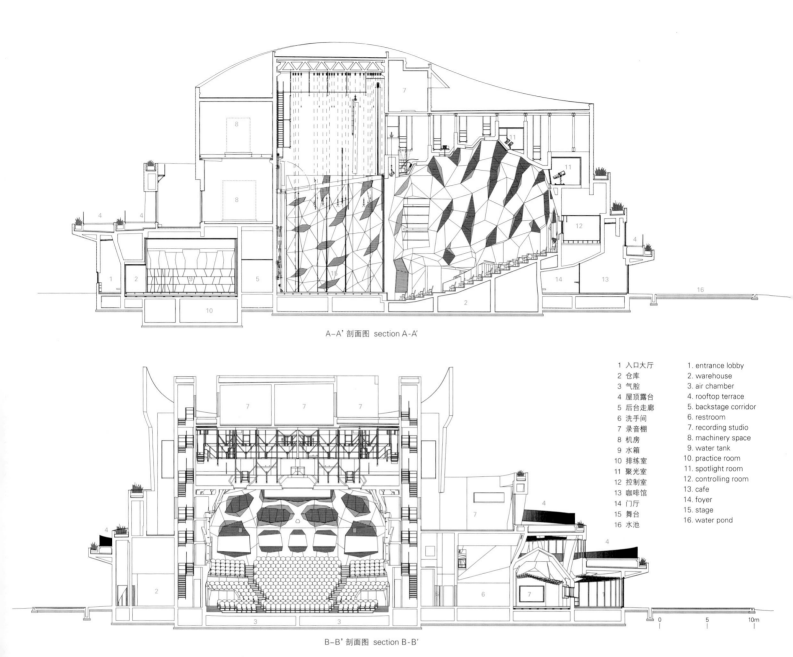

A-A' 剖面图 section A-A'

B-B' 剖面图 section B-B'

1 入口大厅	1. entrance lobby
2 仓库	2. warehouse
3 气腔	3. air chamber
4 屋顶露台	4. rooftop terrace
5 后台走廊	5. backstage corridor
6 洗手间	6. restroom
7 录音棚	7. recording studio
8 机房	8. machinery space
9 水箱	9. water tank
10 排练室	10. practice room
11 聚光室	11. spotlight room
12 控制室	12. controlling room
13 咖啡馆	13. cafe
14 门厅	14. foyer
15 舞台	15. stage
16 水池	16. water pond

CIDAM农业经营研发中心

Landa Arquitectos

坐落在墨西哥米却肯州的农业经营研发中心被设计成高科技的标志，将推动真正的科学型园区的创建和互动。

该中心是总体规划的一部分，主要涵盖行政办公室、调研室、创业园、实验室、图书馆、会议中心，并设有一个庞大的中庭用以展示研究成果。建筑最终成为园区的科学活动集散地及运营中心。

建筑通过模块化的重复性结构设计成为一个灵活的空间，使实验室进行了细分，既满足了空间要求，又具有基础设施的功能。东南至西北方向依次主要设有科研实验室、图书馆、会议中心和多功能中心，以及展示中庭。建筑的主体结构采用混凝土框架，混凝土框架之间的间隔为2.44m，净高达22m。这一设计所创建的空间极具灵活性，而且空间也宽敞。V字形屋顶是框架的最高部分，允许自然光进入。光线经过过滤，在运营区和主要中庭创造了充分的可利用空间。

建筑以其清晰的构造、丰富的空间成为莫雷利亚城市风景的里程碑。建筑结构的稳定性与屋顶空间带来的明亮感形成鲜明对比，使整个建筑既充满活力，又稳如磐石。纵列的圆柱所产生的阴影使建筑动态感十足。

总的来说，建筑塑造了一个鲜活的形象，既能唤醒参观者回忆中的情感，又能突出使用者严谨审慎的创新态度。

CIDAM

The center for the development and investigation of agribusiness in the state of Michoacán was designed to be an icon of high technology, which will promote the creation and interaction of a true scientific community.

This facility forms part of a master plan, destined to host administrative offices, investigation facilities, business incubators, laboratories, library, convention center and a great atrium designed to exhibit the work which is done there. This building results as an integrator of scientific activities and center of operations for the campus.

The building was conceived to be a flexible space, by means of a modular and repetitive structure, which enable the subdivisions of the laboratories, in accordance with the specific requirements

项目名称：CIDAM
地点：Morelia, Michoacán, Mexico
建筑师：Agustin Landa Vertiz
项目团队：Rolando Martinez, Manuel Martinez
甲方：Michoacán State Government
面积：4,200m²
竣工时间：2011.10
摄影师：
courtesy of the architect - p.152, p.157
©Jorge Vertiz(courtesy of the architect) - p.154~155, p.158, p.159, p.160, p.161

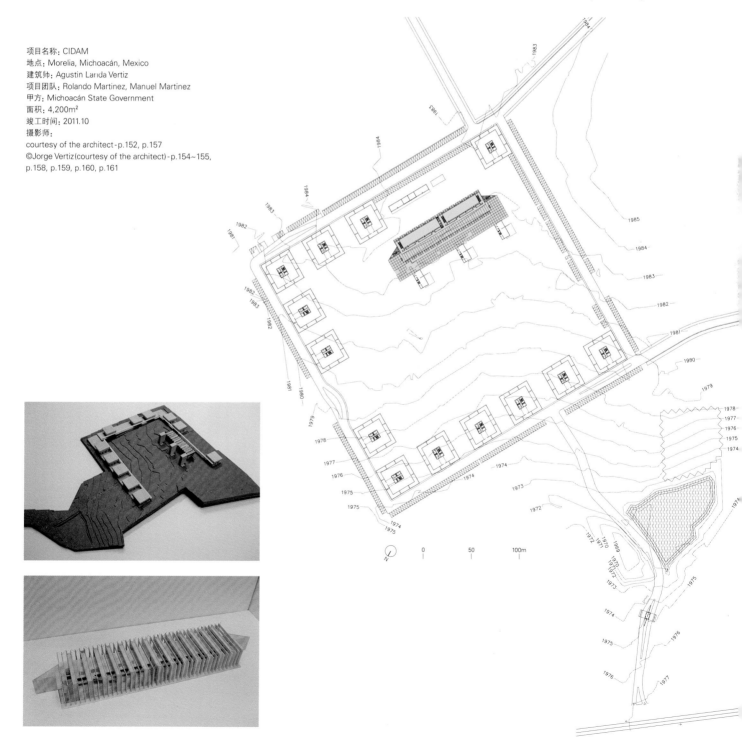

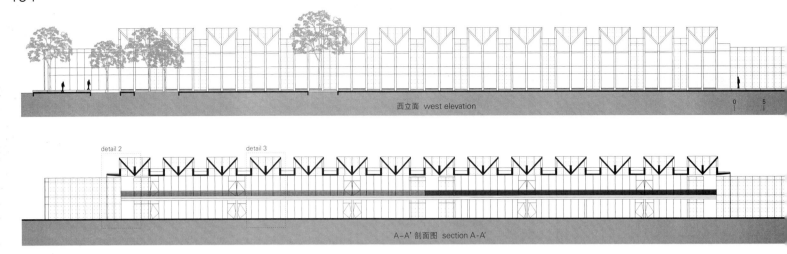

西立面 west elevation

A-A' 剖面图 section A-A'

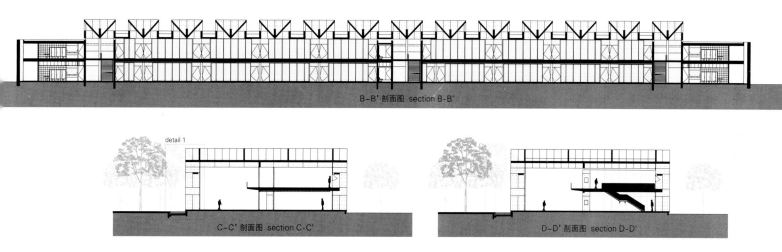

B-B' 剖面图 section B-B'

C-C' 剖面图 section C-C'

D-D' 剖面图 section D-D'

as space and infrastructure. It concentrates on the research laboratories, library, convention and multiuse center, and the exhibition atrium while counting with a southeast-northwest orientation.

The principal structure of the building is based on concrete frames, 2.44 meters apart which allows a clearing of 22 meters. This arrangement creates a space with great flexibility and ample use. In the highest part of this construction, the ceiling in V-shaped form, allows natural light to come in, light which is filtered and sieved, producing a well harnessed space, both in its operational area as in its principal atrium.

With its tectonic clarity and its richness in space, the building presents itself as a milestone in the urban landscape of Morelia. Dynamic and at the same time firm, the building's structure gives solidity, while contrasting with the lightness achieved by the ceiling's clearing. The column array creates shadows, which generate a building dynamic in nature.

As a whole, the facility evokes a fresh image that incites memorable emotions to its visitors, while at the same time, projecting sophistication and commitment of innovation to the people who occupy it.

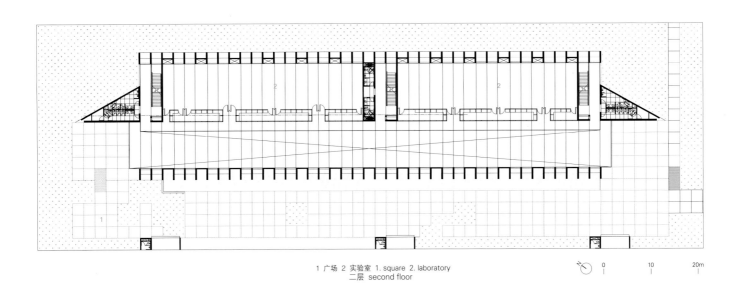

1 广场 2 实验室 1. square 2. laboratory
二层 second floor

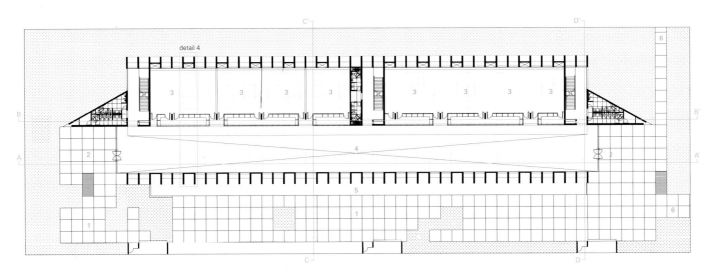

1 广场 2 入口通道 3 实验室 4 大厅 5 水池 6 停车场入口
1. square 2. access 3. laboratory 4. hall 5. water mirror 6. access to parking
一层 first floor

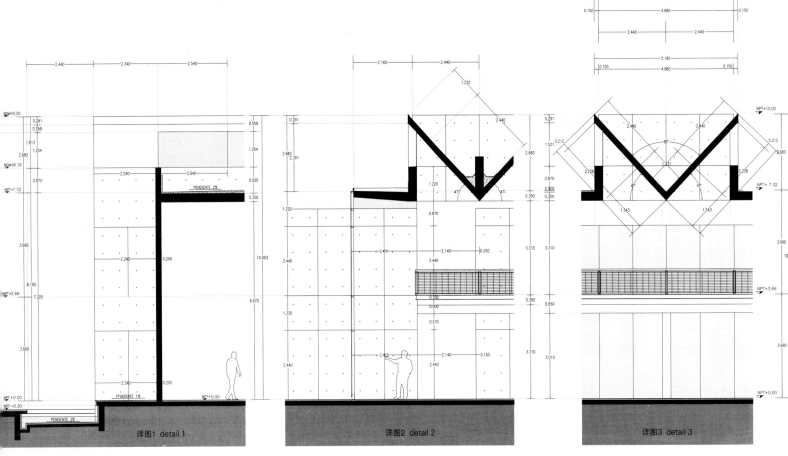

详图1 detail 1

详图2 detail 2

详图3 detail 3

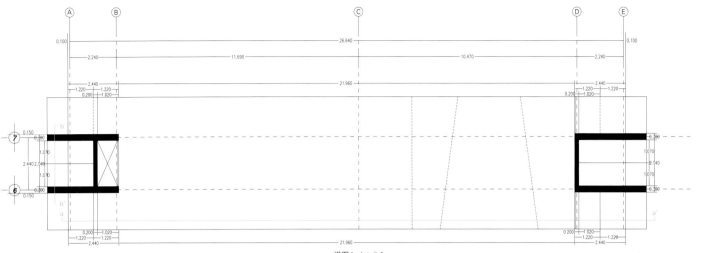

详图4 detail 4

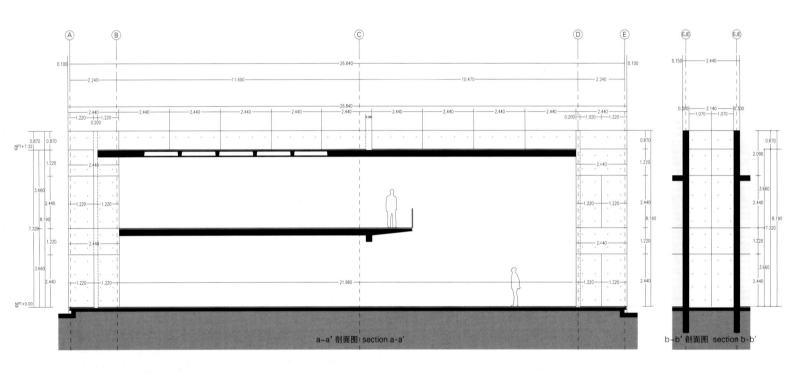

a-a' 剖面图 section a-a' b-b' 剖面图 section b-b'

艺术广场
Brasil Arquitetura

有些建筑项目在宽敞开放的空间中或是有利的环境下显得格外引人注目,即使从远处看也显而易见。其他一些项目则相反,需要与不利的环境、最小的空间、长形地块的窄小楔形空间,以及现有建筑之间的断瓦残垣相适应。在此开发的项目,其特征受到以上种种因素的制约。艺术广场正属于后者的范畴,正如阿尔瓦罗·西扎所说,不应仅仅将空间理解为一种有形物体,还应该是带有兴趣冲突的、具有张力的空间,其特征是未得到充分利用,甚至遭到了废弃,这些特点都要加以考虑。如果一方面艺术广场需要满足与音乐和舞蹈艺术相关的多种新功能的规划要求,那么另一方面,它也不得不清晰地、革新性地与具有紧张生活和稳固邻里关系的现有物理环境和空间环境相呼应。

在圣保罗市中心,这个实际场所三面临街,由一系列与城市街区的中部相连的地块组成。这种环境是错误的城市化建设造成的后果,这种建设方式总得服从于地块理念和私有产业的逻辑。这样就造成了大片未得到充分利用的或者空置的空间。它们被废弃,被遗忘,等待着再次成为城市的兴趣所在。

同时该空间展现了一个特殊的环境,因为其周围的人文因素具有多样性和生命力,各社会阶层混杂,充满了大城市典型的冲突与张力,彼此共生,寻求宽容。简单地说,这是一个富有都市气息的空间。

该项目的方案丰富而又复杂,除包含贯穿于整个综合体、和平共处的公共用途之外,还着重建设音乐和舞蹈活动区。

在常驻表演艺术团体的模块中容纳了多家专业性的艺术组织:市交响管弦乐团、试验储备管弦乐团、抒情唱诗班、圣保罗唱诗班、市芭蕾公司和市弦乐四重奏乐团,这个模块化的体量面对Formosa大街,并合并了

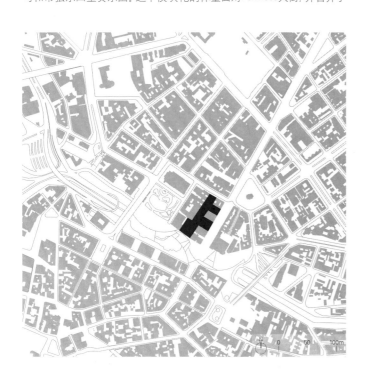

以前开罗电影院的外立面。

学校和公共使用模块容纳了教育和公共空间——市音乐学校、市舞蹈学校、餐馆和公共空间。该模块占据着自São João大街和Conselheiro Crispiniano大街地面上抬高的空间。

由于对最初的建设地点进行了研究，因而要将原先的音乐学校修复成一个音乐厅和一个展览场馆，代表这项工程的标志物。新建筑主要位于基地的边界，而且在很大程度上都是从地面提升起来的，故而创建了开放空间和大面积的流通空间。这样就形成了一个广场空间，整个工程也因此而得名。

这些具有历史意义的建筑既是实体记录，也是象征性的记录，是19世纪末、20世纪初的城市建设中保留下来的老建筑。它们通过全方位的修复和改造而具有了新的用途。在被纳入这项规划之后，这些建筑与其周围建筑关系的变得松散，从而获得了新的意义。

The Square of Arts

Some architectural projects are dominant in large open spaces, in favoured conditions and visible from a distance. Other projects need to adapt to adverse conditions, minimal spaces, small wedges of long plots, leftovers between existing constructions, where the parameters for developing the project are dictated by these factors. The Square of Arts is part of the latter category. To understand the place not only as a physical object, as Alvaro Siza says, but as a space of tension, with conflicts of interest, characterised by underuse or even abandonment, all this counts. If on the one hand the Square of Arts project has to account for the demands of a programme of various new functions, related to the arts of music and dance, it also has to clearly and transformatively respond to an existing physical and spatial situation with an intense life and a strongly present neighbourhood.

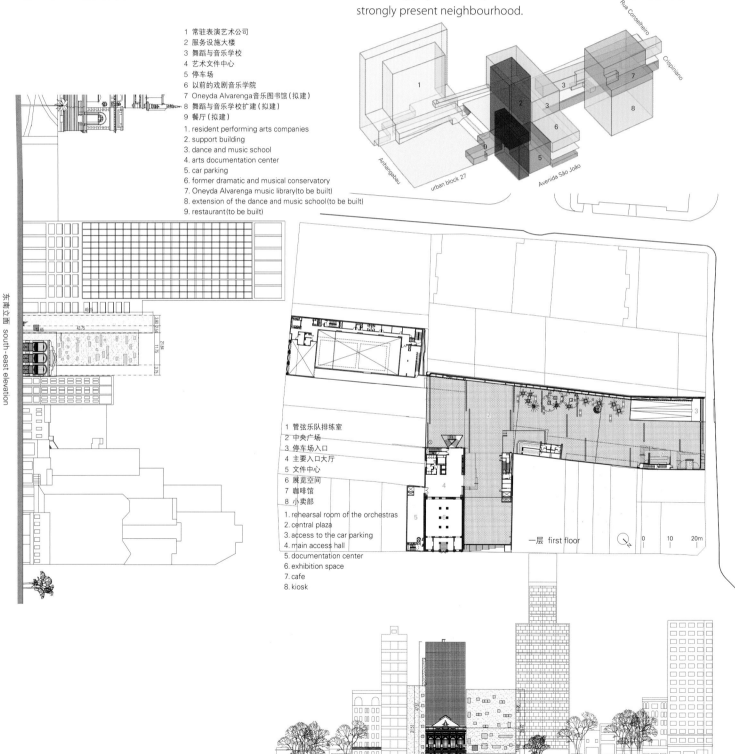

1 常驻表演艺术公司
2 服务设施大楼
3 舞蹈与音乐学校
4 艺术文件中心
5 停车场
6 以前的戏剧音乐学院
7 Oneyda Alvarenga音乐图书馆（拟建）
8 舞蹈与音乐学校扩建（拟建）
9 餐厅（拟建）

1. resident performing arts companies
2. support building
3. dance and music school
4. arts documentation center
5. car parking
6. former dramatic and musical conservatory
7. Oneyda Alvarenga music library(to be built)
8. extension of the dance and music school(to be built)
9. restaurant(to be built)

东南立面 south-east elevation

1 管弦乐队排练室
2 中央广场
3 停车场入口
4 主要入口大厅
5 文件中心
6 展览空间
7 咖啡馆
8 小卖部

1. rehearsal room of the orchestras
2. central plaza
3. access to the car parking
4. main access hall
5. documentation center
6. exhibition space
7. cafe
8. kiosk

一层 first floor

东北立面 north-east elevation

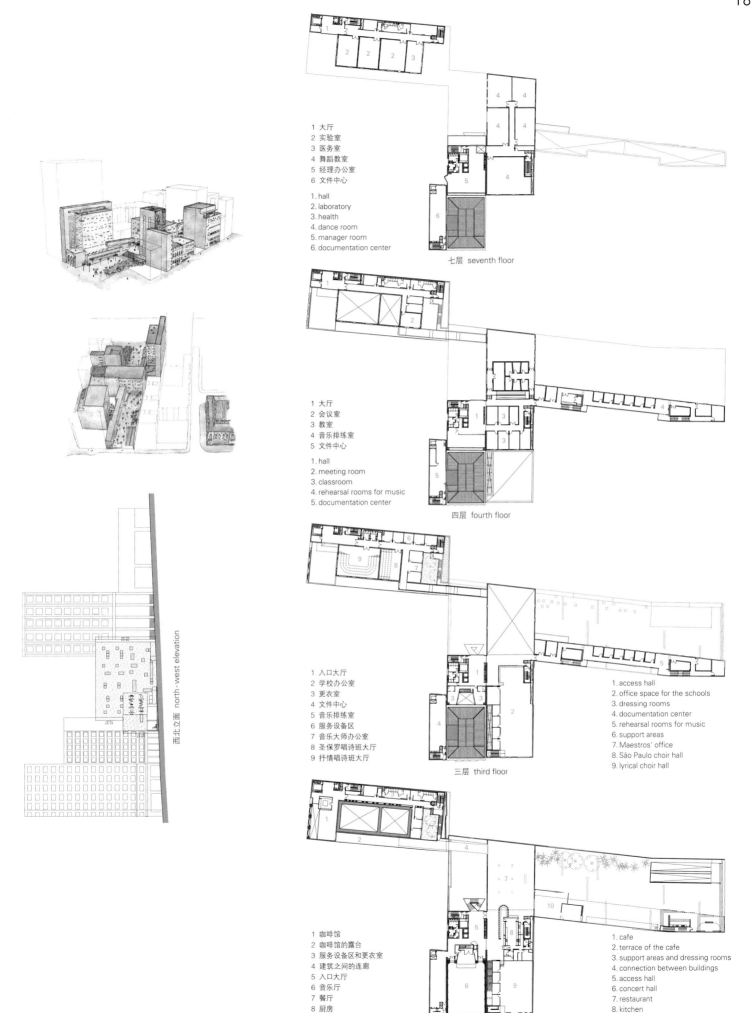

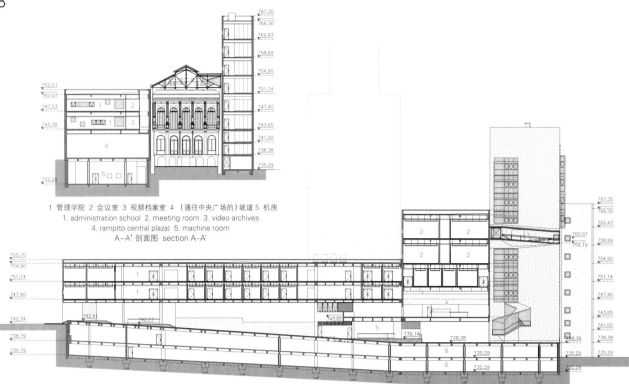

1 管理学院 2 会议室 3 视频档案室 4（通往中央广场的）坡道 5 机房
1. administration school 2. meeting room 3. video archives
4. ramp(to central plaza) 5. machine room
A-A' 剖面图 section A-A'

1 音乐排练室 2 舞蹈教室 3（通往博物馆的）坡道 4 餐厅 5 小吃店 6 停车场
1. music rehearsal rooms 2. dance room 3. ramp(to museum) 4. restaurant 5. snack bar 6. parking
B-B' 剖面图 section B-B'

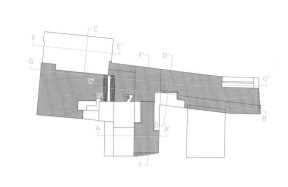

1 实验室 2 档案室
1. laboratory 2. archives
C-C' 剖面图 section C-C'

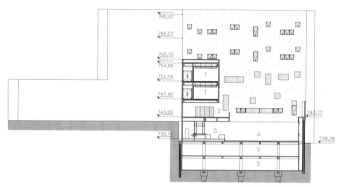

1 钢琴教室 2 阳台 3 吧台 4（通往中央广场的）坡道 5 停车场
1. piano room 2. balcony 3. bar 4. ramp(to central plaza) 5. parking
D-D' 剖面图 section D-D'

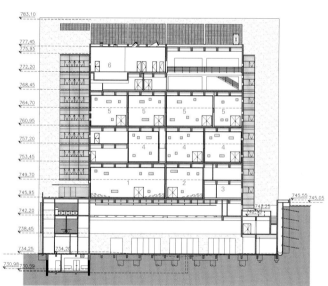

1 抒情唱诗班大厅 2 圣保罗唱诗班大厅 3 音乐大师办公室 4 实验室 5 舞蹈教室 6 水池
1. lyrical choir hall 2. São Paulo choir hall 3. Maestros' office
4. laboratory 5. dance room 6. pool
E-E' 剖面图 section E-E'

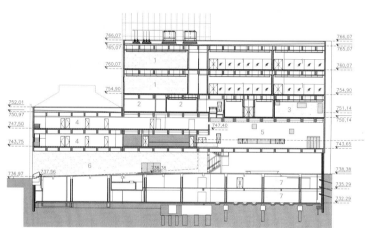

1 舞蹈教室 2 教室 3 打击乐器教室 4 学校办公室 5 餐厅 6 博物馆入口 7 停车场
1. dance room 2. class room 3. percussion room 4. office space for the schools
5. restaurant 6. access to museum 7. parking
F-F' 剖面图 section F-F'

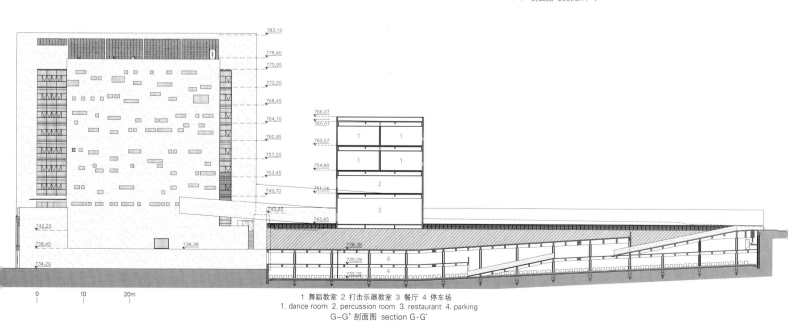

1 舞蹈教室 2 打击乐器教室 3 餐厅 4 停车场
1. dance room 2. percussion room 3. restaurant 4. parking
G-G' 剖面图 section G-G'

The physical place, in the centre of São Paulo, is made up of a series of plots that are connected in the middle of the urban block and have fronts to three streets. This situation is a result of the mistakes of an urbanism that was always subordinated to the idea of the plot, the logic of private property. It is an accumulation of underused or vacant spaces, abandoned, forgotten, awaiting to be of interest to the city once again.

At the same time, the place presents a privileged situation in view of its surrounding humanity, being full of diversity, vitality, a mixture of social classes, conflicts and tensions typical of a large city, living together and the search for tolerance. Shortly, it is a place rich in urbanity.

The project has a rich and complex programme with a focus on musical and dance activities, besides public uses of coexistence, which permeate the entire complex.

The module of the Resident Performing Arts companies houses the professional bodies; the Municipal Symphonic Orchestra, the Experimental Repertory Orchestra, the Lyrical Choir, the São Paulo Choir, the City Ballet Company and the Municipal String Quartet. The module faces rua Formosa(Anhangabaú) and incorporates the facade of the former Cairo Cinema.

The module of the schools and public uses accommodates educational and common spaces – the Municipal Music School, the Municipal Dance School, a restaurant and common space. The module occupies volumes that lift up from the ground on avenida São João and rua Conselheiro Crispiniano.

Since the initial site study, the former Conservatory, restored and converted into a concert hall and an exhibition space, represented the anchor for the project. The new buildings are mainly positioned along the boundaries of the site and, to a large degree, lifted off the ground. Thus, it was possible to create open spaces and generous circulation areas, resulting in the plaza which gives the project its name.

These historic buildings are physical and symbolic records, remains of the city of the end of the 19th and the beginning of the 20th century. Restored in all aspects and converted for new uses, they will sustain a life to be invented. Incorporated them into the project, they became unconfined from neighbouring constructions and gained new meanings.

项目名称：Praça das Artes 地点：São Paulo, Brazil
建筑师：Brasil Arquitetura
项目团队：Anne Dieterich, Beatriz Marques de Oliveira, Felipe Zene, Fred Meyer, Gabriel Grinspum, Gabriel Mendonça, Victor Gurgel, Pedro Del Guerra, Vinícius Spira
助理建筑师：André Carvalho, Júlio Tarragó, Laura Ferraz
合作方：Cícero Ferraz Cruz, Fabiana Fernandes Paiva, Anselmo Turazzi, Carol Silva Moreira
施工文件制作团队：Apiacás Arquitetos /Yuri Faustinoni, Elcio Yokoyama, Ingrid Taets
结构设计：FTOyamada / 照明设计：Ricardo Heder
基础设计：Infraestrutura / 声效与布景：Acústica & Sônica
景观设计：Raul Pereira Paisagismo
用地面积：7,210m² / 总建筑面积：28,500m²
施工时间：2009.1—2012.12
摄影师：
©Leonardo Finotti(courtesy of the architect)-p.162~163, p.170~171, p.175
©Nelson Kon(courtesy of the architect)-p.166~167, p.168, p.172~173, p.174, p.176~177

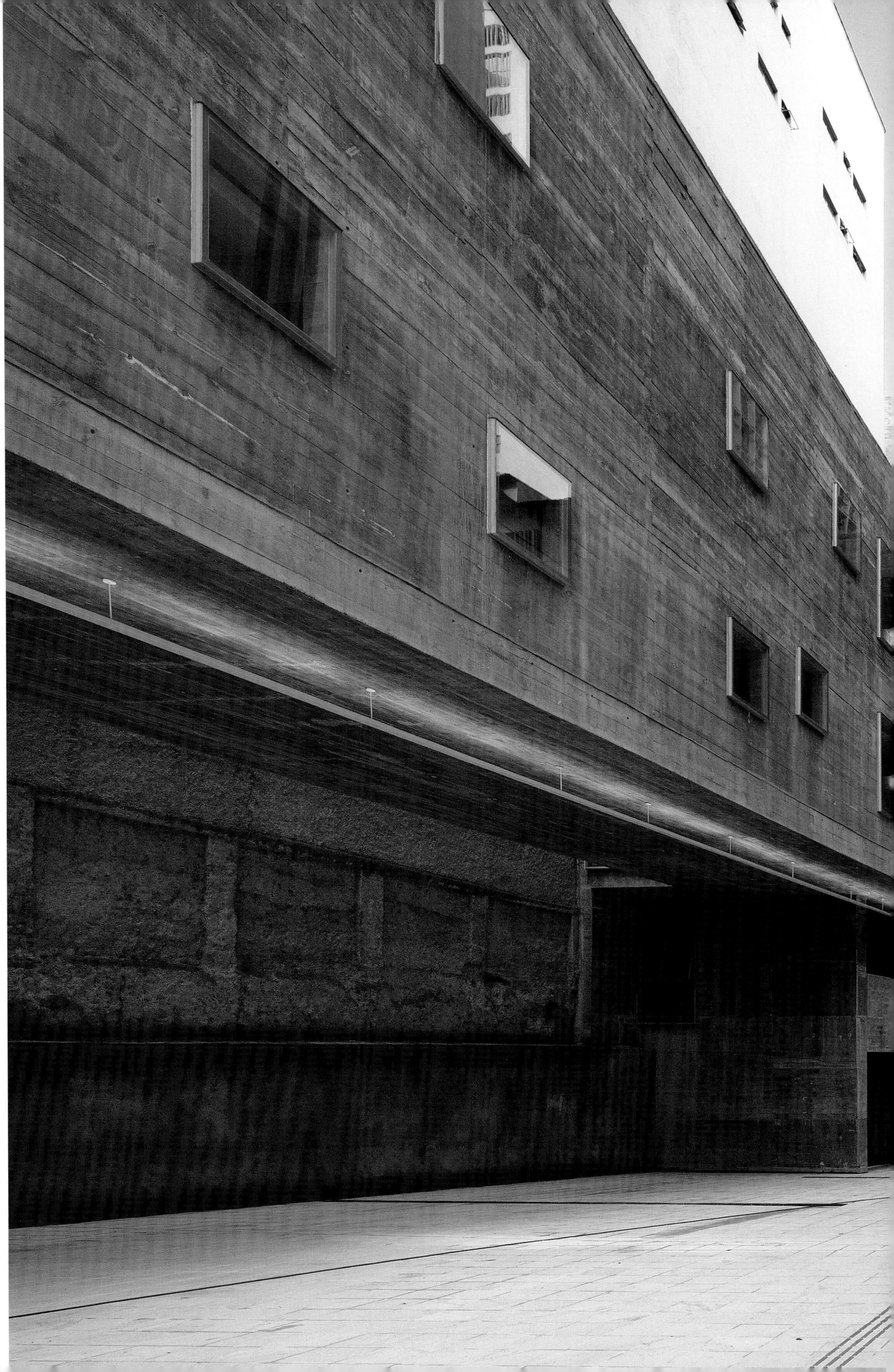

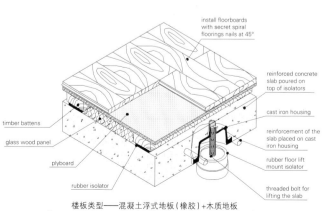

楼板类型——混凝土浮式地板（橡胶）+木质地板
floor type _ concrete floating floor(rubber) + wooden flooring

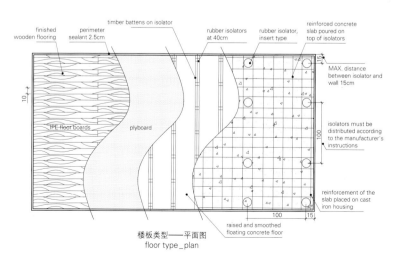

楼板类型——平面图
floor type _ plan

比斯开新文化馆和图书馆

aq4 arquitectura

室内木工: Carpintería Larrondo S.L. Loiu
建筑面积: 3,480m²

OKE是位于西班牙比斯开省老矿区的新奥图埃利亚文化馆。矿区内不能再开采出硫酸铁。这片曾经锈迹斑斑的沙丘如今都融入到混凝土结构之中，在原地堆起了一组笨重的多边形涂漆金属工艺品。矿区的所有过往都已被清除得毫无痕迹。

建筑师了解地下基础设施的规模，知道它们如何出现，也知道这一切是如何被生锈的面漆所污染的。所选用的材料必须能够唤醒人们对这片土地的记忆。建筑师希望建造一座仿佛已长久留存于此的建筑。

建筑与Catalina Gibaja大街的奥图埃利亚主路相垂直，连接了该镇最具特色的公共空间——Otxartaga广场。为了让OKE更引人注目，并且通过周密的设计手法能使其傲然独立于场地之中，建筑师在设计上格外注重其可见性，力争从各个方位都能看到，无论是前后左右，还是哪个犄角旮旯，或者更远的地方。

地块的坡度明显，有利于定义两个不同的入口，每个入口所通往的空间功能都不一样。两个空间都能单独组织安排各种活动。Catalina Gibaja大街（水平高度0.00米）一侧的入口与文化大楼的活动区相连接。该区适于各种活动：运动、宣传、会议，等等。另外，还可以通过6.6m水平高度的人行道一侧的入口进入图书馆：这里有阅览室，人们安静、专注地在这里看书。穿越整个建筑即可完成对其两大功能的感知。

建筑师在3.35m的水平高度创建了一个新的户外空间。这是一个作为公共区域的中间广场，有意远离了周围的活动与嘈杂。从此处可进入多功能室，该房间可独立于大楼单独使用。即使大楼关闭，活动亦可继续进行。停车场入口是对通往Otxartaga广场的旧入口改造而成的。这种方式略微地隔离开了进入大楼和公共空间的步行入口。

建筑的外形与此地起伏的地势紧密相关，极好地定义了该空间：以各种角度倾斜的表面覆盖着原有的广场，并从视觉上跳跃到了山谷的其他角落，述说着抑或是猜想着奥图埃利亚市民想要欣赏的其他城市空间。从山谷看过来，高低错落的屋顶就像是另一种立面，其设计全部根据内部空间的需求得出。

在原有的户外空间与不同高度的活动区相连接的地方，实现了一次新的城市之旅：从通道到图书馆、从通道到广场、从广场到多功能室，再从街道到展厅。

上空间被定义成布满了不同跨度的连续隔间，可根据实际功能需求灵活使用。

这种灵活的布局让文化馆既能提供预期之中的用途，又能满足意料之外的需求。

项目名称: New Culture House and Library in Vizcaya, Ortuella
地点: Calle Catalina Gibaja 10. 48530 Ortuella. Vizcaya. Spain
建筑师: Ibon Bilbao, Jordi Campos, Caterina Figuerola, Carlos Gelpí
参与设计建筑师: Iván Pena, Alberto Berga, Javier González, Albert Duque, María Ruiz, Mónica Molas, Alessandra Sirianni, Marta Milà
结构设计: Bernuz-Fernández arquitectos, Jorge Tejedor, Javier Marta
室内木工: Carpintería Larrondo S.L. Loiu
甲方: Ortuella City Hall
用地面积: 930m²
建筑面积: 3,480m²
造价: EUR 3,348,656
施工时间: 2008—2011
摄影师: ©Adrià Goulà (courtesy of the architect)

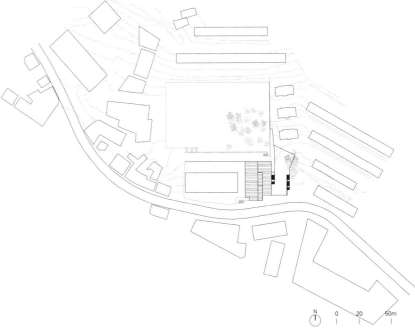

New Culture House and Library in Vizcaya

OKE is the new Culture House of Ortuella in the old mining area of Vizcaya, Spain. There is no extraction of iron sulphate any more. The valleys that previously received rusty sand dunes contained within concrete structures, today pile clumsy groups of lacquered metal crafts in polygons. Everything has been erased, and there are not tracks.

We knew the scale of the underground infrastructure, how they emerged, how everything was dirtied with a rusty varnish. The material must be the one recalling the memory of the site. We wanted to build something that was there long time before. The building is placed perpendicular to the main route of Ortuella, Catalina Gibaja St., connecting it to Otxartaga Plaza, the most characteristic public space in the village. OKE is able to look in both sides, obliquely and as far as possible in order to be seen and been able to see aloof in a precise approach ploy. The strong gradient in the plot favours the definition of two different accesses that organise the two independent uses defining the program. Both can work separately arranging schedules or activities one apart from the other in moment and epoch.

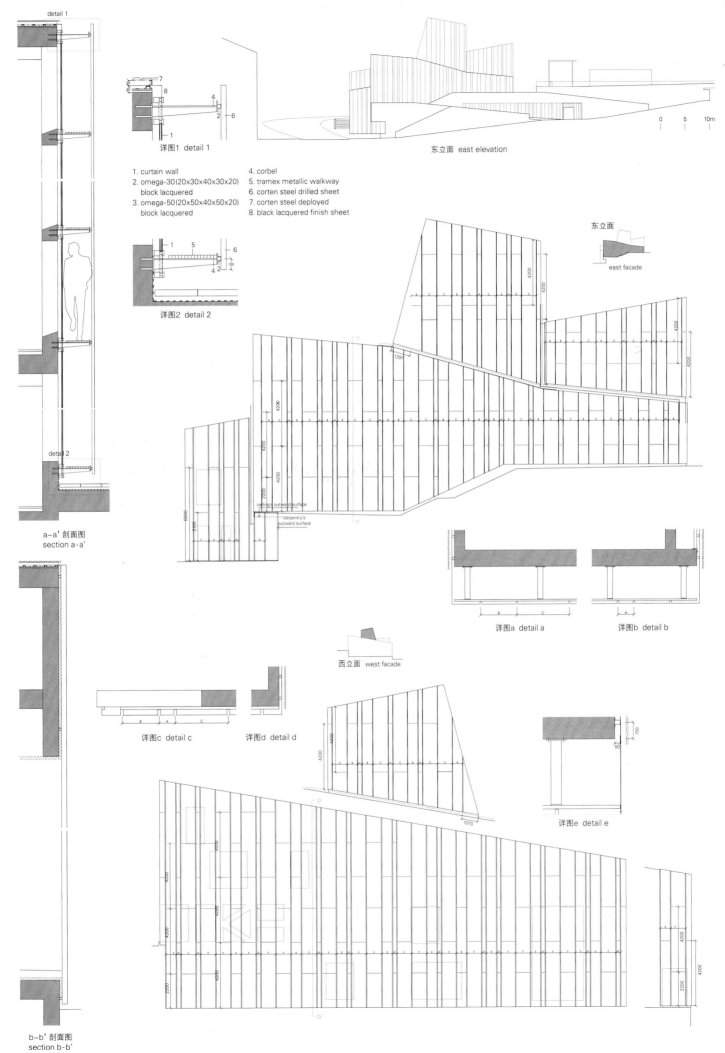

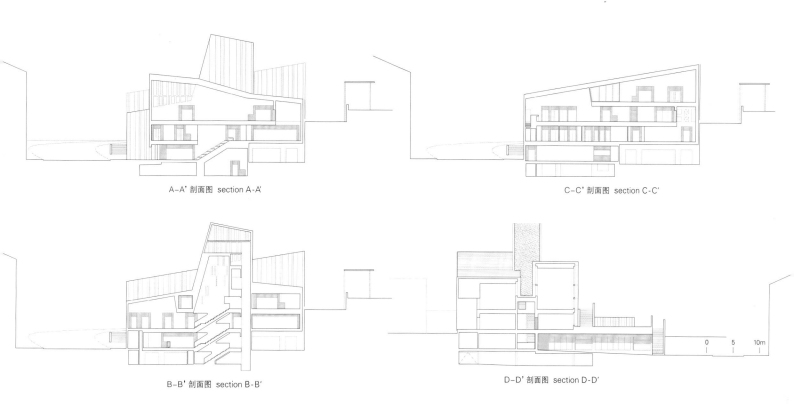

A-A' 剖面图 section A-A'   C-C' 剖面图 section C-C'

B-B' 剖面图 section B-B'   D-D' 剖面图 section D-D'

1 停车场	7 看台	13 多功能室
2 音乐教室	8 行政管理区	14 媒体室
3 储藏室	9 市议员办公室	15 报刊图书馆
4 展览室	10 会议室	16 小组研究室
5 门岗	11 活动室	17 单人研究室
6 门厅	12 休息室	

1. parking	7. stands	13. multipurpose room
2. music room	8. administrative	14. mediatheque
3. storage	9. city councillor office	15. newspaper & magazine library
4. exposition room	10. meeting room	16. group study room
5. post of caretaker	11. room	17. individual study room
6. lobby	12. foyer	

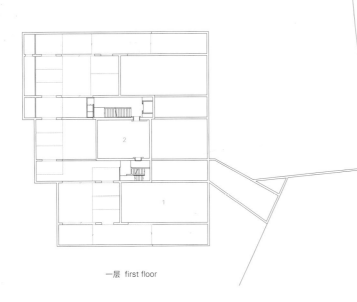

一层 first floor

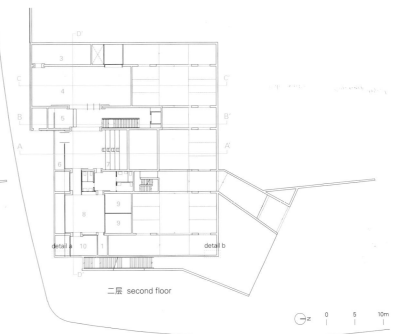

二层 second floor

The entrance from Catalina Gibaja St (level 0.00m) is directly connected with Culture House activities: movement, circulation, events etc.. On the other hand, we can access to the library from a pedestrian passage in level 6.60m: room, silence, concentration. Going over the building completes both program perception.

As a result of the building a new outdoor space is created at 3.35m high. It is an intermediate plaza meant to be a public area apart from the surrounding activity and noise. From this level it is possible to enter the multi-purpose room, the one which can be used independently from the rest of the building and where activities could be ongoing even if the rest of the culture house is closed.

The access to the parking is produced reusing the entrance to the old one from Otxartaga Plaza. In this way the pedestrian access to the building and the public space is slightly interrupted.

The volume that defines the building is shaped in response to the different topographic levels that define the place: tilt covers collect existing square level and visually jump to the other part of the valley, announce, suspect the "other part" that Ortuella would like to enjoy. Looking from the valley roofs are shown as another facade and those are solved as required.

A new urban journey is created where the outdoor existing spaces connect with the different activity levels: walk-library, walk-plaza, plaza-multipurpose room, street-exposition room.

The void is defined – place full of different width span succession where the program is reaching space.

A flexible organization is allowed when some of the spans house services that provide both, expected and unexpected uses from the culture house. aq4 arquitectura

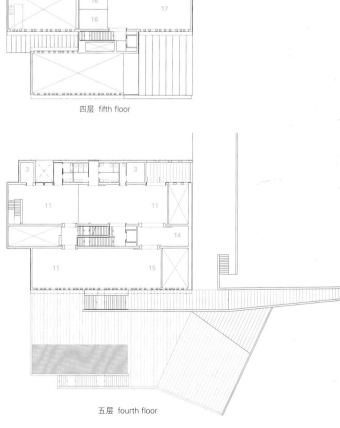

四层 fifth floor

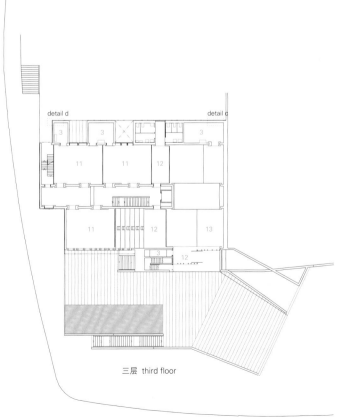

三层 third floor

五层 fourth floor

>>82
Spektrum Arkitekter
Is a young architectural practice leaded by two partners Sofie Willems[left] and Joan Raun Nielsen[right]. Works with urban planning, building and landscape design and ties these disciplines inextricably together. Aims to create vibrant, surprising, inspiring and sensuous spaces contributing to a better quality of life. Strives for a design based on qualities of a specific location, also taking the social context of the site into account. Seeks to involve local people in each project and highly value their needs and desires. Their projects seek to interact with the users, to arouse their curiosity and to invite them to play and explore.

Tom Van Malderen
His activities stretch from the traditional architectural practice to the field of architectural theory which he explores through writing, installations and lectures. After obtaining a master in Architecture at LUCA, Brussels(1997) he worked for Atelier Lucien Kroll in Belgium and in different positions at architecture project, both in the UK and Malta. Lectured at the University of Aix-en-Province in France and the Canterbury University College of Creative Arts in the UK. Contributes to several magazines and publications, and sits on the board of the NGO Kinemastik for the promotion of short film.

Jaap Dawson
Graduated from Cornell University with a Bachelor of Arts in English in 1971. Afterward he earned a Doctor's Degree in Education from Teachers College, Columbia University in 1979. And also received a Ingenieurs diploma(Master of Science) in Architecture from Technische Universiteit Delft in 1988. Currently he delivers a lecture of Architectural Composition in Technische Universiteit Delft and acts as architect, writer, and editor in Delft, the Netherlands since 1988.

>>116
DE-SO Architecture
Was founded in 2005 by Francois Defrain[right] and Olivier Souquet[left]. The founders first started collaborating on projects in 1997 when they took part in several international competitions. Is an award-winning office that is renowned in France and abroad. Specialises in public buildings, housing, urban studies and urban design. Has offices in both France and Vietnam. The Paris studio comprises 15 people of different nationalities; while the office in Ho Chi Minh City employs four Vietnamese architects. Has a broad network of architecture and engineering contacts in both France and Vietnam. The Ho Chi Minh office primarily responds to projects that focus on identity, and sustainable and

Diego Terna
Received a degree in architecture from the Politecnico di Milano and has worked for Stefano Boeri and Italo Rota. Has been working as critic and collaborating with several international magazines and webzines as editor of architecture sections. In 2012, he founded an architectural office, Quinzii Terna together with his partner Chiara Quinzii. Currently is an assistant professor of Politecnico di Milano and runs his personal blog L'architettura immaginata(diegoterna. wordpress.com).

>>12
Scenic Architecture
Zhu Xiaofeng, who is the founder of Scenic Architecture, received a bachelor degree in Architecture from Shenzhen University and his master's degree of architecture from Harvard University Graduate School of Design. In 2004, he established the office based in Shanghai. Believes that the spirit of architecture exists in how people perceive the basics of nature and living. Also trusts that architecture of 21st century shall not only respond to human needs, but also act as positive media between human and environment. Uses architecture to explore how space and time stimulate and absorb each other, and how to establish balanced and dynamic relevance among human, nature and society.

>>152
Landa Arquitectos
Agustín Landa was born in Mexico City in 1951 and graduated from Iberoamericana University(1976) and Oxford Brooks University(1979). Has contributed significantly to the renovation of architecture in Monterrey and to the transformation of the city's landscape. Over the past twenty years, the firm has introduced new typologies to the city, created urban landmarks and innovative public spaces, and contributed to the definition of a local architectural language that values local, industrial materials, efficient plans and construction processes and structural rigor. This language is rooted in Mexican modern architecture and in the history of Monterrey as a city that grew together with its glass, steel and concrete factories.

>>22
Romera y Ruiz Arquitectos
Was founded in 1999 in Las Palmas de Gran Canaria. Their works have been awarded, exhibited and published in various national and international media.
Pedro Romera García[first] has been a professor at the School of Architecture of Las Palmas since 1999, and is a member of the Research Heritage and Landscape Project.
Ángela Ruiz Martínez[second] also has been a professor at the School of Architecture of Las Palmas, general coordinator of the Second Biennial of Art, Architecture and Landscape of the Canary Islands, and member of the Historical Heritage Commissions of Gran Canaria.

Andreas Marx
Andreas Marx's expertise is based on social sciences and urban issues. A first degree in Sociology and Political Science at LMU Munich(2011) strengthened his knowledge of social interaction and the insights of social order. Following a part-time job as researcher for the Institute for Social Science Research, a strong interest for cities emerged and he continued studying at the UVA (Amsterdam 2012-13) - master degree in Urban Sociology. This led to a better comprehension of urban issues in modern cities. His Master Thesis deals with the perception of space and place - in detail with the occurrence and logic of a central daily food market in Munich. His academic focus lies on the question, how one may use the social sciences skills and architecture to improve future city development. Though he recently undertakes a Master degree in Urbanism - Landscape and City at TUM

>>36

SO Architecture
Was established in 2007 by Shachar Lulav and Oded Rozenkier. Shachar Lulav studied interior design in Holon Design Academy and architecture in Vizo Art Academy. Oded Rozenkier studied at Technion-Israel Institute of Technology and at National Superior School of Architecture at Paris-La Villette. They have won several prizes and have been mentioned in several architectural competitions in Israel and abroad.

>>138

Chiaki Arai Urban and Architecture Design
Chiaki Arai was born in Shimane, Japan in 1948 and received first place diploma from Tokyo City University in 1971. Took Advanced Course in University of Pensylvania under Louise I. Kahn in 1973 and worked for the him in the following year. Founded his own office in 1980 and received the UNESCO Asia-Pacific Heritage Awards in 2010 and the International Architecture Awards in 2011. Currently teaches at Tokyo City University.

>>74

Rafael De La-Hoz Arquitectos
Rafael de La-Hoz was born in Córdoba, Spain in 1955. Graduated from the Higher Technical College of Architecture of Madrid, and went on to obtain an MDI Masters from the Polytechnic University of Madrid(UPM). Directs his own architectural studio and works in Spain, Portugal, Poland, Romania, Hungary, and the United Arab Emirates. Is a visiting professor at several universities, and a member of the Editorial Council of COAM architectural magazine.

>>128

Mateo Arquitectura
Is an architectural office based in Barcelona, Spain. Josep Lluís Mateo has studied at the ETSAB and received Ph.D. with cum laude at the Polytechnic University of Catalonia(UPC). Was an editor of the architectural magazine and guest lecturer at many academic institutions. Seeks to connect the practice of construction with research and development in both intellectual and programmatic terms. Currently is a professor of Architecture and Design at the Swiss Federal Institute of Technology(ETH), Zurich.

>>52

Arquitectura 911sc
Is an independent practice based in Mexico City founded in 2002 by Saidee Springall and Jose Castillo, committed to architecture, urban design and planning projects. Has designed and built extensively for both public and private clients in housing, cultural, educational, infrastructure and retail programs.

>>52

Fernanda Canales
Graduated from the University of Iberoamericana(UIA). Received a master's degree in Theory and Criticism from the Polytechnic University of Cataluña(UPC) and Ph.D. in Practice Project from ETSAM. Has taught at the workshop of the UIA and Max Cetto at UNAM and the CMAS in conjunction with the UPC. Has worked as teacher in various exchange programs with universities like ETSAB, besides being jury in competitions and biennials.

>>28

Silvester Fuller
Is an architectural studio based in Sydney, Australia. Was established in 2008 by Jad Silvester[bottom] and Penny Fuller[top].
"Everything is Important." This attitude drives them to find better solutions through the exploration of multiple possibilities and resolve every aspect of each design challenge. Their prior international experiences have led to the creation of their partner offices' global network, giving reach and broader experience to all the projects. They design diverse range of projects in scale from product design through to multi-use tower developments and masterplans. Work across all sectors from cultural, residential and commercial, through retail, public and institutional.

>>88
Kashef Mahboob Chowdhury/Urbana
Kashef Mahboob Chowdhury is the principal of the firm. He was born in Dhaka and studied in Bangladesh and Kuwait. Received a B.Arch from the Bangladesh University of Engineering and Technology(BUET) in 1995. Has worked as professional photographer and has held seven solo exhibitions. Has been a visiting faculty and a juror in final year critic in several universities in Dhaka. His works find root in history with strong emphasis on climate, materials and context-both natural and human.

>>178
aq4 Arquitectura
Started their career with collaboration in architecture workshop "arquitecturaquatre" in 1996 and established architecture aq4 in 2003. Is being led by 4 architects of Ibon Bilbao España, Jordi Campos Garcia, Caterina Figuerola Tomàs and Carlos Gelpí Almirall. They often give lectures and participate in workshops and conferences at institutions and universities around the world. Their works are regularly exhibited and published worldwide and receives international awards.

>>100
b720 Arquitectos
Is an international practice, located in over a dozen different countries. Is operated globally depending on the location of projects. They have offices in Barcelona, Madrid, São Paulo and Porto Alegre. Fermín Vázquez was born in Madrid, 1961 and founded b720 Arquitectos with Ana Bassat in 1997. Graduated from ETSAM and did his Ph.D. at ETSAB. Has taught at ETSAB, University of Barcelona, the École d'Architecture et de Paysage in Bordeaux and at the European University of Madrid.

>>162
Brasil Arquitetura
Was founded in 1979. The practice is run by partners Francisco de Paiva Fanucci and Marcelo Carvalho Ferraz, graduates of the College of Architecture and Urban Design of the University of Sao Paulo(FAU USP). Has developed projects for diverse programs: museums, residential buildings, private houses, boutiques, restaurants, and community and cultural centers. Additionally, the firm is involved in furniture and exhibition design. Currently, Francisco Fanucci and Marcelo Ferraz teach at the Escola da Cidade in São Paulo.

C3, Issue 2014.5

All Rights Reserved. Authorized translation from the Korean-English language edition published by C3 Publishing Co., Seoul.

© 2014大连理工大学出版社
著作权合同登记06-2014年第123号
版权所有·侵权必究

图书在版编目(CIP)数据

都市与社区：汉英对照 / 韩国C3出版公社编；
时真妹等译. —大连：大连理工大学出版社，2014.8
(C3建筑立场系列丛书)
书名原文：C3 Community and the City
ISBN 978-7-5611-9365-5

Ⅰ. ①都… Ⅱ. ①韩… ②时… Ⅲ. ①社区－建筑设计－汉、英 Ⅳ. ①TU984.12

中国版本图书馆CIP数据核字(2014)第164907号

出版发行：大连理工大学出版社
　　　　　（地址：大连市软件园路80号　邮编：116023）
印　　刷：上海锦良印刷厂
幅面尺寸：225mm×300mm
印　　张：12
出版时间：2014年8月第1版
印刷时间：2014年8月第1次印刷
出 版 人：金英伟
统　　筹：房　磊
责任编辑：张昕焱
封面设计：王志峰
责任校对：赵姗姗

书　　号：978-7-5611-9365-5
定　　价：228.00元

发　行：0411-84708842
传　真：0411-84701466
E-mail: dutp@dutp.cn
URL: http://www.dutp.cn